Springer Natural Hazards

The Springer Natural Hazards series seeks to publish a broad portfolio of scientific books, aiming at researchers, students, and everyone interested in Natural Hazard research. The series includes peer-reviewed monographs, edited volumes, textbooks, and conference proceedings. It covers all categories of hazards such as atmospheric/climatological/oceanographic hazards, storms, tsunamis, floods, avalanches, landslides, erosion, earthquakes, volcanoes, and welcomes book proposals on topics like risk assessment, risk management, and mitigation of hazards, and related subjects.

More information about this series at http://www.springer.com/series/10179

Yingbin Zhang

Earthquake-Induced Landslides

Initiation and Run-out Analysis by Considering Vertical Seismic Loading, Tension Failure and the Trampoline Effect

Science Press
Beijing

Springer

Yingbin Zhang
Department of Geotechnical Engineering
Southwest Jiaotong University
Chengdu, Sichuan
China

ISSN 2365-0656 ISSN 2365-0664 (electronic)
Springer Natural Hazards
ISBN 978-981-10-2934-9 ISBN 978-981-10-2935-6 (eBook)
DOI 10.1007/978-981-10-2935-6

Jointly published with Science Press, Beijing

The print edition is not for sale in China Mainland. Customers from China Mainland please order the print book from: Science Press, Beijing

Library of Congress Control Number: 2017941469

Printed on acid-free paper

This Springer imprint is published by Springer Nature
The registered company is Springer Nature Singapore Pte Ltd.
The registered company address is: 152 Beach Road, #21-01/04 Gateway East, Singapore 189721, Singapore

Preface

A large number of landslides can be induced by a strong earthquake, and they can cause very serious damage to both lives and properties. Much research has been carried out, and a series of countermeasures have been developed to reduce the adverse effects of landslide disasters. In the planning and design of countermeasures, it is necessary to analyze the stability of a slope and movement behaviors of a potential landslide under seismic loadings. Therefore, this study focuses on analysis of (1) slope stability and (2) landslide movement behaviors.

In the study of slope stability, tension-shear failure mechanism of seismic slope is presented. Most solutions for slope stability analysis are derived just based on shear failure mode, although tension failure always exist in earthquake-induced landslides. Thus, how to analyze slope stability according to both shear and tension modes is an important problem. This problem is discussed in detail, and both analytical and numerical methods are presented to solve them.

In the study of landslide movement behaviors, long run-out, one of the major behaviors of earthquake-induced landslides, is discussed. Since the mechanism is still not clear, the multiple acceleration model (MAM) is extended on the basis of the so-called trampoline effect. Additionally, a practical numerical simulation program is developed to clarify movement behaviors of earthquake-induced landslides in practical simulations.

Combining these two aspects, slope stability and landslide movement, is an effective way to study the whole process of earthquake-induced landslides. Discontinuous deformation analysis (DDA) is a discrete numerical method that was developed for computing large deformation and large displacement in discontinuous block system. However, the original DDA cannot simulate the cohesive material accurately. This monograph presents an extended DDA that can accurately simulate the failure behavior of cohesive-frictional material by using an additional evaluation of edge-to-edge contact. Several examples are performed to illustrate the accuracy, and a real landslide case is studied to clarify the whole process of earthquake-induced landslide.

This Monograph has received financial support from the National Natural Science Foundation of China (41672286 and 51408511); Science & Technology Department of Sichuan Province (2017JQ0042); Japan Society for the Promotion of Science (JSPS) KAKENHI (16F16056); School of Civil Engineering, Southwest Jiaotong University. The financial supports are gratefully acknowledged.

Chengdu, China Yingbin Zhang

Contents

Notations and Acronyms

2-D	Two-dimension
3-D	Three-dimension
A	Amplitude of input wave, gal, m/s^2
a	Empirical constant parameter of rock mass
$a_{(+)}$(t)	Upward acceleration of ground vibration, m/s^2
$a_{(-)}$(t)	Downward acceleration of ground vibration, m/s^2
a_c	Critical acceleration, m/s^2
a_E	East acceleration of ground, m/s^2
a_h	Horizontal pseudo-static acceleration, m/s^2
a_H	Horizontal acceleration of ground, m/s^2
ALOS	Advanced Land Observing Satellite
a_N	North acceleration of ground, m/s^2
a_{para}	Ground acceleration parallel to the slide surface, m/s^2
a_{perp}	Ground acceleration perpendicular to the slide surface, m/s^2
a_v	Vertical pseudo-static acceleration, m/s^2
a_V	Vertical acceleration of ground, m/s^2
BEM	Boundary element method
c	Cohesion of slope material, kPa
C	Damping matrix
c^{critical}	Critical cohesion of slope material, kPa
CEA	China earthquake administration
COSMOS	Consortium of organizations for strong-motion observation systems
C_{s}	Speed of s-wave propagation through material
CS	Contact state
c_s	The minimum wave velocity of seismic wave
D	Length of CC
	Matrixes of displacement
	Blast damage and stress relaxation factor of rock mass
	Fraction of critical damping

$\dot{D}$	Matrixes of velocity
$\ddot{D}$	Matrixes of acceleration
DDA	Discontinuous deformation analysis
DEM	Discrete element method
	Digital elevation model
DF	Distance to fault
D_i	Deformation variable of block i
D_m	Run-out distance, km
D_{max}	Maximum displacement, m
D_n	Newmark displacements, m
D_{res}	Residual displacement, m
DZ	Dangerous zone
E	Modulus of elasticity, Pa
EQ	Earthquake
EW	East–west
f	Characteristic frequency, Hz
	Maximum frequency of incident wave, Hz
F	Forcing matrix of DDA equations
$\tilde{F}$	Effective force matrix of DDA equations
FDM	Finite difference method
FEM	Finite element method
f_h	Horizontal pseudo-static force, N
F_i	Load of block i, N
f_i	Non-dimensional function related to vertical direction
f_i'	Non-dimensional function related to horizontal direction
FLAC^{3D}	Fast Lagrangian analysis of continua in three-dimension
FOS	Factor of safety
FOS$_S$	Factor of safety based on shear-only failure mechanism
FOS$_{TS}$	Factor of safety based on tensile-shear failure mechanism
f_{shear}	Shear failure function
$f_{tension}$	Tension failure function
f_v	Vertical pseudo-static force, N
G	Shear modulus, Pa
g	Gravitational acceleration, 9.81m/s^2
g_0	Normal contact spring stiffness in DDA
g_1	Step time in DDA
g_2	Assumed maximum displacement ratio in DDA
gg	Static or dynamic parameter in DDA
GIS	Geographic information systems
H	Height of slope, m
h_m	Drop height of collapse mass under multi-movements, m
h_s	Drop height of collapse mass under sliding movement, m
I_a	Arias Intensity
ICOLD	International commission of large dams

$\dot{\varepsilon}_{ij}$	Strain tensor
λ_{ij}	Non-negative scalar function
JAXA	Japan aerospace exploration agent
K	Stiffness matrix in DDA equations
$\tilde{K}$	Effective stiffness matrix of DDA equations
k_c	Critical seismic coefficient
k_h	Horizontal pseudo-static coefficient
k_i^*	Multiple friction coefficient
K_{ii}	Material properties of block i
K_{ij}	Contact between blocks i and j
k_n	Stiffness of the added normal springs
K-NET	Kyoshin net
k_s	Stiffness of the added shear springs in DDA
k_v	Vertical pseudo-static coefficient
k_{vpr}	Velocity preserve rate
k_{vtr}	Velocity transmittable ratio
L	Length of top of slope failure, m
l	Contact length of two blocks, m
LAM	Limit analysis method
LEM	Limit equilibrium method
LMSF	Longmenshan fault
m	Mass of rigid block, kg
M	Earthquake magnitude
$\bar{\varphi}_{\mathrm{m}}$	Apparent angle under multi-movements, °
MAM	Multiplex Acceleration Model
m_b	Material constant in Hoek–Brown failure criterion
MMI	Modified Mercalli Intensity
N	Supporting force, N
n	Number of calculation steps
NGA Database	Data from Pacific Earthquake Engineering Research Center
NS	North–south
PGA	Peak ground acceleration, m/s^2
PGA_H	Horizontal Peak Ground Acceleration, m/s^2
PGA_{rock}	Peak ground acceleration of bedrock, m/s^2
PGA_V	Peak Ground Acceleration of Vertical Component, m/s^2
$r(\theta)$	Radius of spiral related to angle θ, m
r_0	Initial values of the radius of the spiral, rad
	Rigid rotation around block center, rad
RBSM	Rigid block spring method
RD	Relative Difference
R-F	Rossi–Forel earthquake intensity scale
s	Parameter of rock mass in Hoek–Brown failure criterion
$\dot{W}_{seismic}$	External rate of seismic work, J
$\dot{D}_{shear}$	Shear energy dissipation, J

$\dot{D}_{tension}$	Tension energy dissipation, J
SRF	Strength reduction factor
SR-FEM	Strength reduction-finite element method
SSI_{max}	Maximum values of shear strain increment
SSR	Union of Soviet Socialist Republic
STA	Station
T	Displacement transformation matrix in DDA equations
U	Displacement of block in DDA
u_0	Translations of block centroid along the x-axis
v_0	Translations of block centroid along the y-axis
UD	Up-down
v	Velocity, m/s
$v(\theta)$	Rotational velocity, rad/s
V'	Velocity of large block after collision, m/s
v_{out}	Velocity of small block after collision, m/s
V	Velocity of large block before collision, m/s
v_{in}	Velocity of small block before collision, m/s
v_k^{res}	Residual velocity in kth collision, m/s
v_s	Input shear particle velocity, m/s
W	Gravity, kN
	Weight of sliding block, kN
$\dot{W}_{weight}$	External rate of weight work, J
Z	Depth of tension crack, m
α	Slope angle, °
	Inclined upper slope angle, °
a_L	Local damping coefficient
β	Slope angle, °
β'	Sliding out angle, °
γ	Unit weight, kN/m^3
$\Delta_{\text{a-b}}$	Varying percentage of the sliding displacement
Δl	Maximum mesh size of simulation model, m
ε_{x}	X-normal strains of block
ε_y	Y-normal strains of block
γ_{xy}	Shear strains of block
η	Strike angle of slope, °
η_k	Ratio of residual velocity to precollision velocity
θ	Angle of spiral, rad
θ_0	Initial value of the angle of spiral, rad
θ_h	Final value of the angle of spiral, rad
λ	Shortest wavelength of incident wave
$\dot{\lambda}$	Non-negative scalar function
μ	Poisson's ratio
μ_f	Falling friction coefficient
ξ	Ratio of k_v to k_h

ρ	Mass density of soil material, kg/m^3
σ_1	Major principal stress, kPa
$\sigma_1{}'$	Combined major principal stress, kPa
	Maximum effective stresses at failure, kPa
σ_3	Minor principal stress, kPa
$\sigma_3{}'$	Combined minor principal stress, kPa
	Minimum effective stresses at failure, kPa
$\sigma'_{3\max}$	Maximum of minor principal stress, kPa
σ_{ij}	Stress tensor, kPa
σ_s	Stress history, kPa
σ_t	Tension strength, kPa
$\sigma_t{}'$	Reduced tension strength, kPa
τ_s	Shear strength, kPa
φ	Internal friction angle, °
$\bar{\varphi}$	Apparent friction angle or angle of reach
φ'	Reduced internal friction angle, °
χ	Coefficient of residual and maximum displacement
ψ	Dilation angle, °
ω	Angular velocity, rad/s

List of Figures

List of Tables

Chapter 1
Introduction

Abstract This chapter introduces (1) the study background, (2) two main issues in the study of landslides, namely slope stability and landslide movement behaviors, (3) the scope and objectives of this study, and (4) the organization of the monograph.

Keywords Slope Stability · Landslides · Organization

1.1 Background

1.1.1 Landslide

The term '*landslide*' describes the downward and outward movement of slope-forming materials, e.g., rock, soil, artificial fill, or a combination of them, with a wide variety of processes, including flowing, sliding, toppling, falling, or a combination of two or more types of movements according to Glossary of Geology (Jackson 1997) and other references (Varnes 1974; Hutchinson 1988; WP/WLI 1990; Cruden and Varnes 1996; Highland and Bobrowsky 2008; Gokceoglu and Sezer 2009).

Landslide is one of the most serious natural hazards worldwide because it frequently occurs and potentially threats to both life and property (UNEP 1997; EM-DAT 2003). For example, (1) more than 60,000 landslides induced by the 2008 Wenchuan earthquake caused about 20,000 fatalities; (2) landslides on Hispaniola Island in May 2004 caused over 2500 fatalities in Haiti and the Dominican Republic (Nadim et al. 2006). Table 1.1 lists the ranking of major natural hazards by deaths from EM-DAT (2003). It is believed that the number of causalities due to landslides shown in the Table 1.1 is grossly underestimated because the loss figures in the international databases were normally recorded by the primary triggering factor, and not by the hazard that causes the fatalities (Nadim et al. 2006). For example, more than 80,000 deaths are recorded as the Wenchuan earthquake self, while about a quarter of the total deaths were caused by the earthquake-induced landslides (Yin et al. 2009).

Y. Zhang, *Earthquake-Induced Landslides*, Springer Natural Hazards,
DOI 10.1007/978-981-10-2935-6_1

Table 1.1 Ranking of major natural hazards by deaths in EM-DAT (2003) (avalanche is not included in landslides)

Rank	Disaster type	All deaths	Deaths (1992–2001[a])
1	Drought	563,701	277,574
2	Storms	251,384	60,447
3	Floods	170,010	96,507
4	Earthquakes	158,551	77,756
5	Volcanoes	25,050	259
6	Extreme temperature	19,249	10,130
7	Landslides	18,200	9461
8	Wave/surges	3068	2708
9	Wildfires	1046	574
Total		1,211,159	533,416

[a]*2002 IFRC World Disaster Report* (http://www.cred.be/emdat/intro.htm)

Landslides can be induced by several triggers, such as intense or prolonged rainfall, strong earthquakes, rapid snow melting, and a variety of human activities (Guzzetti 2006).

Therefore, it is very important to study the landslide so as to mitigate the potential disaster from landslides.

1.1.2 Earthquake-Induced Landslides

Earthquake is one of the major triggers for landslides. Figure 1.1 shows the relationship between hot spots of global landslides and worldwide large earthquakes. The landslide distribution consists well with the large earthquake distribution.

A strong earthquake can induce a large amount of landslides and cause very serious property damage and human casualties. This phenomenon was recorded at least as early as in ancient China dated back to 1789 BCE (3792 years ago), and in ancient Greece 2385 years ago (Keefer 2002). There have been many reports about very serious damages caused by the earthquake-induced landslides for the last few decades, especially after a series of disastrous earthquake events occurred in recent years. For example, 9272 landslides induced by the 1999 Chi-Chi earthquake (Ms = 7.6) caused 2400 deaths, more than 8000 casualties, and over 10 billion US$ of economic loss in Taiwan (Chang et al. 2005). About 30% of the total fatalities (officially 87,350) were killed by co-seismic landslides induced by the 2005 Kashmir earthquake (Ms = 7.6) (Havenith and Bourdeau 2010). Less than three years later, the 2008 Wenchuan earthquake shocked the Sichuan Province and induced as many as 60,104 landslides (Gorum et al. 2011), which directly caused more than 20,000 deaths (Yin et al. 2009), a quarter of the total deaths, and over one-third of the total lost was caused by the earthquake-induced landslides. Table 1.2 shows earthquakes responsible for triggering landslides.

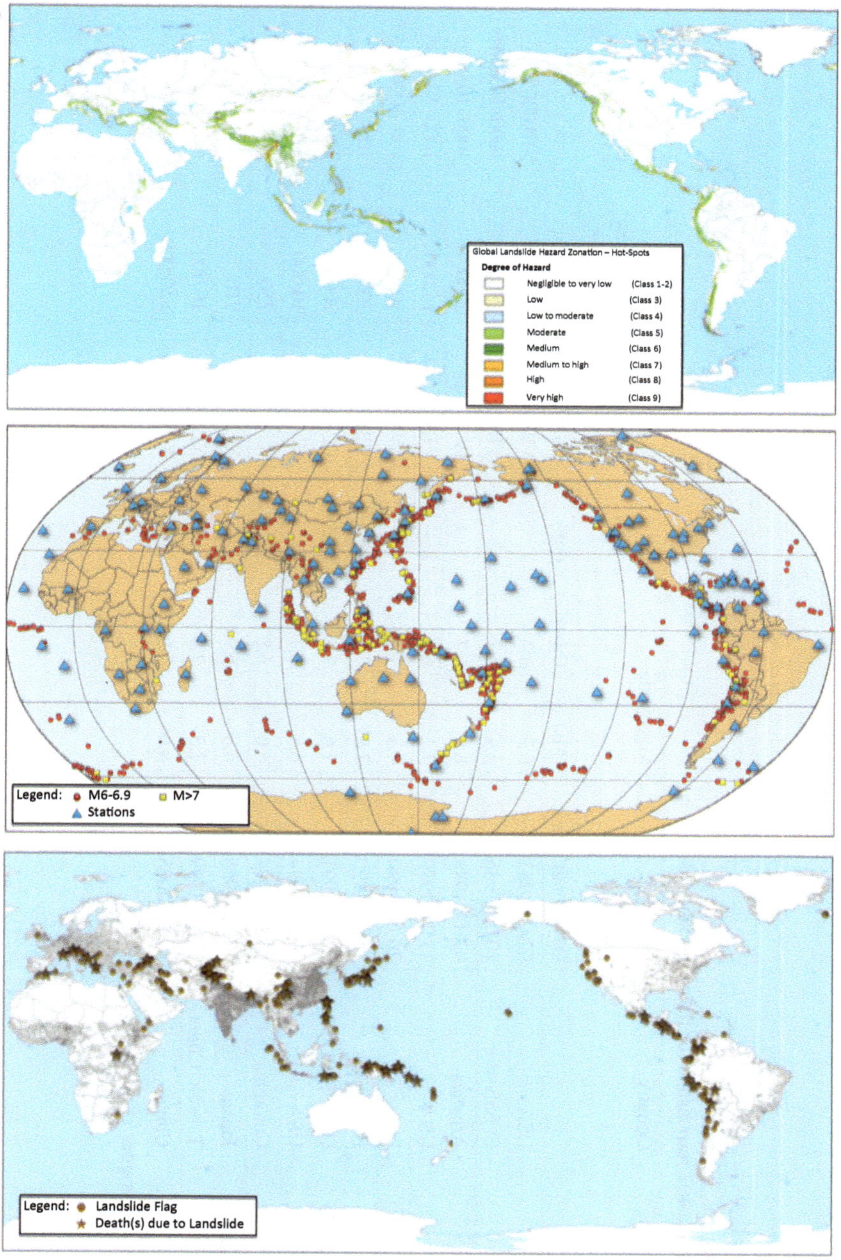

Fig. 1.1 Relationship between **a** hot spots of global landslides (modified from Nadim et al. 2006), **b** worldwide large earthquakes (*red circles*—magnitude 6–6.9, *yellow squares*—magnitude 7 and larger earthquakes, *triangles*—seismographic stations) (from USGS open Fact Sheet (2011). This Fact Sheet is available online at http://pubs.usgs.gov/fs/2011/3021), and **c** epicenters of earthquake-induced landslides from September 1968 to June 2008 (from Marano et al. 2010)

Table 1.2 Earthquakes responsible for triggering landslides

Earthquake	Country	Date	Magnitude		Focal depth	Maximum intensity	Area affected by landslides	Number of slides
		Day/month/year	(Ms)	(Mw)	(km)	(MMI)	(km^2)	
Coalinga	USA	02 05 1983	6.7	6.2	7	VIII	650	>10,000*
San Salvador	El Salvador	10 10 1986	5.4	5.7	12	VIII	380	1000–10,000*
Spitak	Armenia	07 12 1988	6.8	6.7	5	IV	2200	1000–10,000*
Loma Prieta	USA	17 10 1989	7.1	6.9	8	VIII	14,000	1000–10,000*
Manjil	Iran	20 06 1990	7.3	7.4	19	X	1000	100–1000*
Luzon	Philippines	16 07 1990	7.8	7.7	25	VIII	3000	100–1000*
Valle de la Estrella	Costa Rica	22 04 1991	7.6	7.5	21.5	IX	2000	1000–10,000*
Northridge	USA	17 01 1994	6.8	6.7	18	IX	10,000	>10,000*
Paez	Colombia	06 06 1994	6.6	6.8	12	X	250	1000–10,000
Hyogo-Ken Nanbu	Japan	17 01 1995	6.8	6.9	22	X	910	100–1000
Chi-Chi	Taiwan, China	21 09 1999	7.3	7.6	8	XI	10,000	≈10,000
Wenchuan	China	12 05 2008	8.0	7.9	19	XII	>50,000	>60,000

Note *Data from Rodriguez et al. (1999)

Figure 1.2 shows two examples of landslides induced by the 2008 Wenchuan earthquake, the new Beichuan landslide and Wangjiayan landslide. These two landslides destroyed a middle school and part of an old town, respectively, and killed at least 2500 people, in which about one thousand are students. They are two of the most catastrophic landslides caused by the 2008 Wenchuan earthquake. Therefore, study on the earthquake-induced landslide is necessary.

A number of studies on the seismic slope stability analysis and landslide run-out analysis have been carried out, and a series of countermeasures have been developed to mitigate the landslide disasters. However, these studies cannot be used to explain a number of landslides induced by the earthquakes occurred in recent years which have attracted the attentions of geologist and geotechnical engineer since these landslides have unique failure mechanism, complex dynamic process, and large-scale sliding with high speed and long run-out. These characters are quite different from the existing knowledge. It is necessary to develop new analysis methods to give more reasonable explanation for those earthquake-induced landslides with special failure mechanism and unique long run-out.

1.2 Preventive Countermeasures and Two Major Issues

Preventive countermeasures include hard measures, e.g., anchor, pile, and soft measures, e.g., policy, hazard map, warning system. In order to provide useful information for designing or planning preventive countermeasures, we need first to identify the potential landslides as well as their affected areas.

There are two major issues in study on earthquake-induced landslides (Fig. 1.3). The first one is how to evaluate the stability of a slope under seismic loadings, i.e., seismic slope stability. The second one is where and how far it will reach once the failure occurs, i.e., run-out analysis. In view of disaster prevention, the first issue helps us to identify the landslide prone slope. The second issue studies landslide movement behaviors, containing failing path, run-out distance, velocity or energy, and deposit distribution, which provides useful information for setting up preventive structures. In addition, tension-shear failure mechanism and long run-out movement mechanism of potential seismic landslide should be considered in these two major issues, respectively.

Therefore, study of the stability analysis and movement behaviors of a potential landslide under seismic loadings is very important and necessary.

1.3 Scope and Objectives

For the seismic slope stability analysis, there are three general categories methods: (1) factor of safety (FOS)-based pseudo-static methods, (2) displacement-based dynamic sliding block methods, and (3) stress-strain methods. Most of the existed methods take the slope failure as a completely shear failure while ignoring the

Fig. 1.2 Comparison of Beichuan County before (**a**) and after (**b**) the Wenchuan earthquake. The earthquake induced two catastrophic landslides, **c** new Beichuan middle school landslide, it destroyed the school and killed 907 people, most of them were students, and **d** Wangjiayan landslide, it buried the old town and killed at least 1600 people (*Photographs* are from Tang) identify the landslide prone slopes as well as their affected areas first

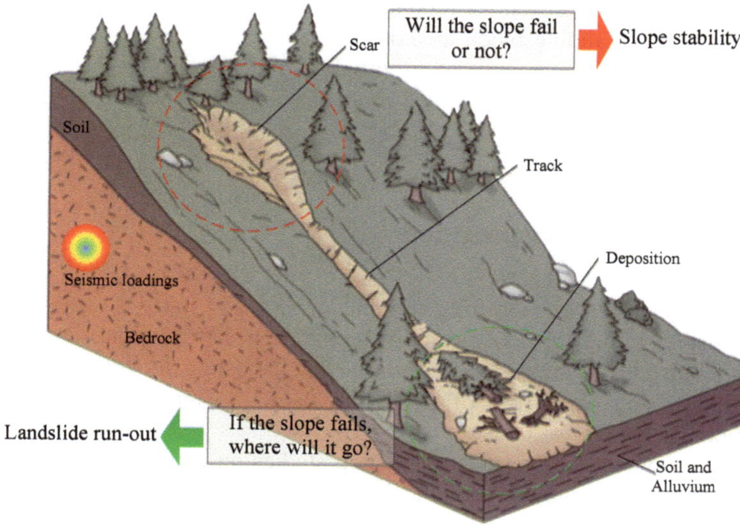

Fig. 1.3 Two main issues in earthquake-induced landslides

tension failure. However, the tension failure has been observed in many earthquake-induced landsides, top surface of potential landslides and shaking table tests.

For landslide run-out analysis, there are four general categories methods: (1) experiment methods, (2) empirical methods, (3) analytical methods, and (4) numerical simulation methods. Results of these methods have been used to produce some laws or codes for guiding practice engineering. For example, based on statistical results, twice of slope height is taken as the danger zone in Japan. However, many earthquake-induced landslides travelled much longer than the danger zone.

As mentioned above, the seismic loadings have significant influence on the seismic slope stability and landslide run-out. Thus, some questions are arising: Does the tension failure play an important role in seismic slope stability? Whether the seismic loadings can cause significant tension failure? What about the permanent displacement induced by the seismic loadings? What is the role of seismic loadings on the landslide run-out? In order to answer these questions, both analytical method and numerical simulation method are used to evaluate the effects of seismic loadings on earthquake-induced landslides from two aspects: seismic slope stability and landslide run-out.

The objectives of this study are to solve the following problems in realization of answering the above questions:

(1) To present new methods for evaluating the effect of tension failure on seismic slope stability.
(2) To present new methods for evaluating the effects of seismic loading on earthquake-induced permanent displacement.

(3) To propose new approach for producing a seismic landslide hazard map based on the dynamic sliding block method.
(4) To develop numerical simulation program for simulating the earthquake-induced landslide run-out by considering the trampoline effect of seismic loadings.

1.4 Monograph Organization

The monograph comprises the following chapters.

This chapter introduces (1) the study background, (2) two main issues in the study of landslides, namely slope stability and landslide movement behaviors, (3) the scope and objectives of this study, and (4) the organization of the monograph.

Chapter 2 reviews two aspects of existing studies on earthquake-induced landslides: slope stability analysis and landslide movement simulation. The merits and demerits of each method are stated.

Chapter 3 proposes an approach for how to consider tension failure in addition to shear failure in widely used FLAC^{3D}, a finite difference method, so that stability analysis can also be carried out for a slope with a complex slip surface. It is shown that the safety factors estimated using the numerical methods are almost the same comparing to an analytical method for a homogeneous slope. Additionally, it is shown by a large number of analysis examples that the effect of tension failure on slope stability is significant and the safety factor will be incorrect if tensile failure is ignored in the case of seismic loading.

Chapter 4 presents a long run-out model based on the so-called earthquake-induced trampoline effect and develops a practical numerical simulation program for estimating landslide movement behaviors. The multiple acceleration model (MAM) is derived from mechanism analysis of the earthquake-induced trampoline effect. The results show that the proposed new long run-out model is reasonable and applicable.

Chapter 5 extends the original discontinuous deformation analysis (DDA). A practical numerical simulation program is developed by incorporating the MAM into the extended DDA. After an extreme ground movement with the peak ground acceleration (PGA) of 4000 gal is successfully reproduced, some large-scale landslides induced by the 2008 Sichuan Earthquake are analyzed in practical numerical simulations. The results show that the movement behaviors of earthquake-induced landslides can be analyzed using the numerical simulation program.

Chapter 6 presents a case study to verify the proposed new methods from slope stability analysis to landslide run-out analysis. The Daguangbao landslide, the largest scale landslide induced by the 2008 Sichuan earthquake, is analyzed using a numerical simulation program DDA as well as FLAC^{3D}. The results show that the vertical component of seismic loading may play an important role in both stability analysis and run-out analysis, as larger tension failure and trampoline effects may be induced by the vertical seismic force, which has generally been ignored up to now.

Chapter 7 concludes the results and achievements of the study, and states the problems to be solved in future studies.

References

Chang, K.-J., Taboada, A., Lin, M.-L., & Chen, R.-F. (2005). Analysis of landsliding by earthquake shaking using a block-on-slope thermo-mechanical model: Example of Jiufengershan landslide, central Taiwan. *Engineering Geology, 80*(1–2), 151–163.

Cruden, D. M., & Varnes, D. J. (1996). Landslide types and processes. In A. K. Turner & R. L. Schuster (Eds.,) *Landslides, investigation and mitigation. Transportation Research Board Special Report 247*, (pp. 36–75). Washington D.C.

EM-DAT. (2003). The OFDA/CRED International disaster database. Brussels, Belgium, Universit´e Catholique de Louvain. www.em-dat.net

Gokceoglu, C. & Sezer, E. (2009). A statistical assessment on international landslide literature (1945–2008). *landslides 6*(4), 345–351

Gorum, T., Fan, X., van Westen, C. J., Huang, R. Q., Xu, Q., Tang, C., et al. (2011). Distribution pattern of earthquake-induced landslides triggered by the 12 May 2008 Wenchuan earthquake. *Geomorphology, 133*(3–4), 152–167.

Guzzetti, F. (2006). *Landslide hazard and risk assement*. Perugia: University Bonn.

Havenith, H. B., & Bourdeau, C. (2010). Earthquake-induced landslide hazards in mountain regions. A review of case histories from central asia. *Geologica Belgica, 13*(3), 137–152.

Highland, L. M., & Bobrowsky, Peter. (2008). *The landslide handbook—A guide to understanding landslides*. Reston, Virginia, U.S.: Geological Survey Circular.

Hutchinson, J. N. (1988). General report: Morphological and geotechnical parameters of landslides in relation to geology and hydrology. In *5th international symposium on landslides*, Lausanne (pp. 3–35)

Jackson, J. A. (1997). Glossary of geology. Alexandria, Virginia: American Geological Institute

Keefer, D. K. (2002). Investigating landslides caused by earthquakes—a historical review. *Surveys in Geophysics, 23*, 473–510.

Marano, K. D., Wald, D. J., & Allen, T. I. (2010). Global earthquake casualties due to secondary effects: A quantitative analysis for improving rapid loss analyses. *Natural Hazards, 52*(2), 319–328.

Nadim, F., Kjekstad, O., Peduzzi, P., Herold, C., & Jaedicke, C. (2006). Global landslide and avalanche hotspots. *Landslides, 3*(2), 159–173.

Rodriguez, C. E., & Bommer, J., Chandler, R. J. (1999). Earthquake-induced landslides: 1980–1997. *Soil Dynamics and Earthquake Engineering, 18*, 325–346

UNEP. (1997). *World Atlas of Desertification*. London: Arnold.

Varnes, D. J. (1974). The logic of geological maps, with reference to their interpretation and use for engineering purposes. *US Geological Survey Professional Paper, 837*, 48.

WP/WLI. (1990). A suggested method for reporting a landslide. *International Association Engineering Geology Bulletin, 41*, 5–12

Yin, Y., Wang, F., & Sun, P. (2009). Landslide hazards triggered by the 2008 Wenchuan earthquake, Sichuan China. *Landslides, 6*(2), 139–152.

Chapter 2
Review of Studies on Earthquake-Induced Landslides

Abstract This chapter reviews two aspects of existing studies on earthquake-induced landslides: slope stability analysis and landslide movement simulation. The merits and demerits of each method are stated.

Keywords Review · Slope stability · Landslides · Earthquake-induced landslides

2.1 Introduction

A large number of landslides can be caused by a strong earthquake. Appendix A shows a general list of 99 historical earthquakes causing substantial landsliding in the worldwide since the twentieth century. From Appendix A, a huge number of landslides were induced by the Chi-Chi earthquake in Taiwan (1999), by the Mid-Niigata Prefecture earthquake in Japan (2004), by the Northern Pakistan earthquake (2005), and by the Wenchuan earthquake in China (2008). Most recently the Great East Japan earthquake with magnitude 9.0 also caused a lot of landslides.

Earthquake-induced landslides have been the source of significant damage and loss of people and property. Landslides caused the majority of damage or casualties in many earthquakes. One of the most extreme examples is from the 1970 Peru earthquake, which triggered a huge rock avalanche that buried two cities and killed almost half the 54,000 fatalities (Rodriguez et al. 1999). Another example is, in the 1920 Haiyuan earthquake, numerous large landslides caused widespread devastation to buildings and infrastructure and killed at least 100,000 people, almost half of the total earthquake deaths (Keefer 2000).

Therefore, it is very important to predict the earthquake-induced landslides and to take countermeasures for potential landslides.

The problem of seismic landslide hazards has received considerable attention worldwide in recent years. Some special international symposiums for earthquake-induced landslides have been hold on and some special journal issues or proceedings on this field have been published (Table 2.1).

Y. Zhang, *Earthquake-Induced Landslides*, Springer Natural Hazards,
DOI 10.1007/978-981-10-2935-6_2

Table 2.1 Some special issues on earthquake-induced landslides from international journals

No.	Journal/Press	Issue/data	Name
1	Engineering Geology	2000, vol. 58, issue. 3–4	Special issue from the symposium on landslide hazards in seismically active regions
2	Surveys in Geophysics	2002, vol. 23, issue. 6	Special issue of the symposia on assessment and mitigation of collateral seismic hazards
3	Engineering Geology	2006, vol. 86 issue. 2–3	Landslides induced by earthquake and volcanic activity
4	Engineering Geology	2011, vol. 122, issue. 1–2	The next generation of research on earthquake-induced landslides
5	Springer	2013. Feb.	Earthquake-induced landslides Proceedings of the International Symposium on Earthquake-induced landslides, Kiryu, Japan, 2012

Main topics of earthquake-induced landslides are the following:

1. Investigation of recent and historical earthquake-induced landslides and their impacts so as to produce inventories of historical earthquake-induced landslides.
2. Prediction of potential earthquake-induced landslides, including (i) failure mechanism and stability analysis of seismic slopes, (ii) movement mechanism and behaviours of earthquake-induced landslides, and (iii) Instrumentation and monitoring technologies for potential earthquake-induced landslides or post-earthquake landslides.
3. Preventive countermeasures for earthquake-induced landslides, including (i) Stabilization and disaster mitigation of earthquake-related landslides, (ii) risk assessment and management of earthquake-related landslides, and (iii) hazard map and early warning system for earthquake-related landslides.

This chapter focuses on the prediction of potential earthquake-induced landslides. The prediction of potential landslide can be carried out using detailed geotechnical investigations and stability calculations. (i) Failure mechanism and stability analysis of seismic slopes, i.e., seismic slope stability analysis and (ii) movement mechanism and behaviours of earthquake-induced landslides, i.e., landside run-out analysis are outlined firstly, and then, the merits and demerits of each method are clarified in this chapter.

2.2 Seismic Slope Stability Analysis

Methods developed to date to assess the stability of slopes during earthquakes can be divided into three categories: (1) pseudo-static methods, (2) dynamic sliding block methods, and (3) stress-strain methods. Each of these types of methods has strengths and weaknesses, and each can be appropriately applied in different situations (Jibson 2011).

The Sect. 2.2 reviews these three categories methods and discusses their advantages and limitations.

2.2.1 *Pseudo-Static Methods*

Terzaghi (1950) first presented the pseudo-static method, which is a simple method for evaluating of seismic stability of a slope. It can be applied to natural or man-made slopes based on either analytical method or numerical method. The earthquake force acting on the whole or an element of the slope is represented by a horizontal force and/or a vertical force equal to the product of the gravitation force and a coefficient *k*, called the pseudo-static seismic coefficient as shown in Fig. 2.1.

Thus, the assumed seismic acceleration a is k times the gravitational acceleration g, i.e., $a = kg$. On a potential sliding mass of weight W, the assumed pseudo-static forces will be

$$\begin{aligned} f_h &= \frac{a_h W}{g} = k_h W \\ f_v &= \frac{a_v W}{g} = k_v W \end{aligned} \tag{2.1}$$

where a_h and a_v are horizontal and vertical pseudo-static accelerations, respectively; k_h and k_v are horizontal and vertical pseudo-static coefficients, respectively. The FOS is expressed as the ratio of the resisting force to the driving force, Eq. 2.2.

$$\text{FOS} = \frac{\tau_r}{\tau_d} \tag{2.2}$$

There are three points for the pseudo-static seismic force mentioned: (1) **magnitude**, (2) **direction**, and (3) **acting point**, which should be determined in the application of the pseudo-static methods.

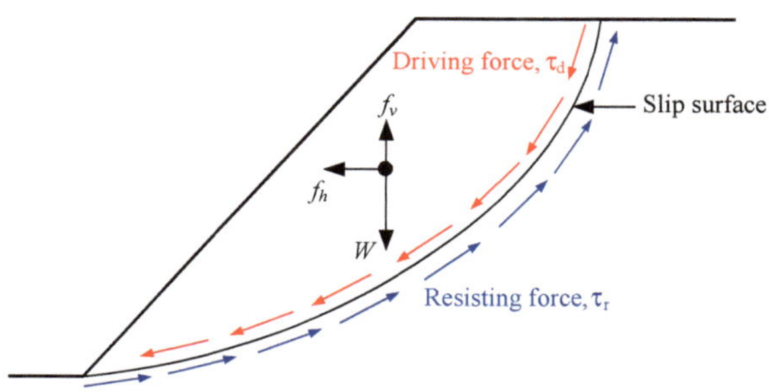

Fig. 2.1 Forces acting on a slope in pseudo-static slope stability analysis

2.2.1.1 Magnitude

From Eq. 2.1, the magnitude of the pseudo-static force is determined by the seismic coefficient. The key problem for the pseudo-static procedure is how to select an appropriate seismic coefficient under an acceptable FOS. There have been studies for determining the most appropriate pseudo-static coefficient by a matter of experience and judgment.

Terzaghi (1950)'s classical paper, probably the earliest recommendations on the values of the seismic coefficient published by a renowned geotechnical engineering or engineering geologist, made the original suggestion to use of k_h = 0.1 for **severe** earthquakes, k_h = 0.2 for **violent** and/or **destructive** earthquakes, and of k_h = 0.5 for **catastrophic** earthquakes.

Makdisi and Seed (1977) recommended a minimum pseudo-static factor of safety of 1.5 based on a soil strength reduction factor of 0.8 and the following values of acceleration associated with two different values of earthquake magnitude *M*. The same values of seismic coefficients for magnitude 6.5 and 8.25 earthquakes are recommended by Seed (1979), but with an acceptable FOS of 1.15.

$$\begin{aligned} a &= 0.1\text{g for } M = 6.5 \text{ (implying } k = 0.1) \\ a &= 0.15\text{g for } M = 8.25 \text{ (implying } k = 0.15) \end{aligned} \qquad (2.3)$$

Seed (1979) gives a summary of the horizontal seismic coefficients and minimum factor of safety value for 14 large dams worldwide in the report published by the International Commission of Large Dams (ICOLD). The horizontal seismic coefficients range from 0.1 to 0.15 and the minimum factors of safety range from 1.0 to 1.5. The Corps of Engineers Manual (EM-1110-2-1902), published in 1982, recommended a seismic coefficient value of 0.1 or 0.15 for regions where earthquake threat is major and great, respectively, and a minimum FOS of 1.0 for all magnitude earthquakes. Historically, values of *k* in the range 0.05–0.15 have been adopted for seismic analysis and design of dams in the USA even in regions of very high seismic susceptibility such as California. In Japan, which also has high seismic susceptibility, typical values of *k* less than 0.2 have generally been adopted (Seed 1979).

Marcuson and Franklin (1983) and Hynes-Griffin and Franklin (1984) related the seismic coefficient value for a dam to the expected PGA at a site. Marcuson and Franklin (1983) recommended a seismic coefficient of 1/3–1/2 of the PGA at the crest of a dam, whereas Hynes-Griffin and Franklin (1984) recommended a seismic coefficient of 1/2 of the PGA of bedrock (PGA_{rock}) with a minimum factor of safety of 1.0 and a 20% reduction in shear strength. Bray and Rathje (1998) also related the seismic coefficient to the PGA of bedrock (0.6 or $0.75PGA_{rock}$). This value is appropriate for seismic stability evaluations of solid-waste landfills, where the allowable levels of deformation are relatively small. Kramer (1996) pointed that although engineering judgment is required for all cases, the criteria of Hynes-Griffin and Franklin (1984) should be appropriate for most slopes.

Pyke (1991) showed that, for an 8.25 magnitude earthquake, if a pseudo-static analysis using a seismic coefficient equal to one-half the peak acceleration yields a factor of safety greater than 1.0, the displacements are likely to be acceptably small. Similarly, for earthquakes with magnitudes 7.5, 7.0, and 6.5, if the seismic coefficients are taken one-third, one-fourth, and one-fifth of the PGA, and the computed factors of safety are greater than 1.0, the displacements are likely to be acceptably small.

Krinitzsky (1993) suggested one-half of PGA to use in an area of low seismicity (peak acceleration < 0.15 g) for the stability of earth embankments. This can be obtained from the peak horizontal motion (mean) from Modified Mercalli Intensity (MMI), magnitude-distance attenuation, and the probability of a 50-year, 90% non-exceedance. However, in an area of moderate to strong seismicity ($0.15\ g \leq PGA \leq 0.40\ g$), PGA is obtained from the peak horizontal motion, from MMI, magnitude-distance attenuation, and probability of 250-year, 90% non-exceedance.

Kavazanjian et al. (1997) suggested a minimum pseudo-static FOS of 1, also based on a soil strength reduction factor of 0.8 and the following values of 'a', expressed as a proportion of the PGA on soil sites: $a = 0.17$ PGA, if dynamic response analysis is also to be performed for the slope or earthquake structure, and $a = 0.5$ PGA, if dynamic response analysis is not to be performed for the slope or earth structure.

Stewart et al. (2003) used the data of Bray and Rathje (1998) to develop an expression for the seismic coefficient in terms of ground motion (PGA, duration) and earthquake parameters (magnitude). These seismic coefficient values generally range from 0.25 to $0.75PGA_{rock}$.

Table 2.2 lists several recommendations for selecting a pseudo-static coefficient. All of these recommended pseudo-static coefficients can fall into two categories: (1) magnitude-based coefficients and (2) peak ground acceleration (PGA)-based coefficients.

In particular, it is recommended that pseudo-static analysis, which provides only a very rough approximation of slope behavior during earthquake shaking, should be used only for preliminary assessments and screening procedures, then followed by more sophisticated analysis (Stewart et al. 2003; Wasowski et al. 2011; Jibson 2011).

2.2.1.2 Direction

It is almost common that only the horizontal acceleration is considered in evaluating the stability and deformation of a slope because the horizontal acceleration is the principal de-stabilizing force that acts on earth structures as well as the principal source of damage observed in earthquakes (Anderson and Kavazanjian 1995).

Table 2.2 Pseudo-static coefficient from several studies

Recommended pseudo-static coefficient (*k/g*)	FOS	Permanent displacement (*D/m*)	Original application	References
0.1 (R-F = IX) 0.2 (R-F = X) 0.5 (R-F = XI)	>1.0	–	Nature or man-made slope	Terzaghi (1950)
0.1 (M = 6.5) 0.15 (M = 8.25)	>1.5	<1	Earth dams	Makdisi and Seed (1977)
0.1 (M = 6.5) 0.15 (M = 8.25)	>1.15	<1	Earth dams	Seed (1979)
(1/3 ~ 1/2) PGA	>1.0	–	–	Marcuson (1981)
1/2 PGA	>1.0	<1	Earth dams	Hyness-Griffin and Franklin (1984)
1/2 PGA (M = 8.25) 1/3 PGA (M = 7.5) 1/4 PGA (M = 7.0) 1/5 PGA (M = 6.5)	>1.0	–	–	Pyke (1991)
0.15	>1.1	–	Dams	CDCDMG (1997)
(0.6 ~ 0.75) PGA_{rock}	>1.0	<0.15 ~ 0.3	Solid-waste landslides	Bray and Rathje (1998)
(0.25 ~ 0.75) PGA_{rock}	>1.0	<0.05 ~ 0.15	Urbanized slopes	Stewart (2003)

Note R-F is Rossi-Forel earthquake intensity scale, IX severe earthquake, X destructive earthquake, XI catastrophic earthquake, M is earthquake magnitude, PGA is peak ground acceleration, in terms of acceleration of gravity

From Fig. 2.1, the horizontal force clearly increases the driving force and decreases the FOS. The vertical pseudo-static force typically has less influence on the FOS than the horizontal pseudo-static force does because the vertical pseudo-static reduces both the driving force and resisting force. As a result, the effects of vertical accelerations are frequently omitted in pseudo-static analysis (Kramer 1996).

Analyses performed by several investigators with an inclined seismic force (i.e., coupled with vertical component of the earthquake force) have shown that the inclination can have a significant influence on the seismic slope stability analysis (Chopra 1966; Ling 1998).

The critical value of the FOS is not necessarily associated with the horizontal direction of the lateral force. In general, a horizontal direction is assumed for the lateral force. There may, however, be situations in which the vertical component of the earthquake force must be taken into consideration.

2.2.1.3 Acting Point

Incorporating the lateral force in an analysis requires a decision as to its point of application. Generally, it is applied at the centre of gravity of a potential sliding mass or of a typical vertical slice in any method of slices. Alternatively, the force may be applied at the level of the base of a slice. Different points of application of pseudo-static force can induce a significant difference in the result. Seed (1979) provided a well-known example, the analysis of Sheffield Dam. In his study, the seismic forces were applied at the base and the center of gravity of each slice, respectively, and the results of factor of safety were 1.21 and 1.32, respectively (Chowdhury et al. 2010).

In summary, pseudo-static method can be simply and directly used to identify the FOS and the critical seismic coefficient k_c, but it cannot simulate the transient dynamic effects of earthquake shaking, because it assumes a constant unidirectional pseudo-static acceleration. In addition, performance of slope is closely related to permanent displacement, but the results of pseudo-static method are difficult to interpret the performance of slope after a seismic event because this method provides no information about permanent displacement. Therefore, in particular, it is recommended that pseudo-static methods should be used only for preliminary assessments and screening procedures, which then should be followed by more sophisticated analysis (Stewart et al. 2003; Wasowski et al. 2011; Jibson 2011).

2.2.2 *Dynamic Sliding Block Methods*

Displacement-based dynamic sliding block method is another alternative approach to evaluate the seismic slope stability, as permanent displacement is a useful index of slope performance, especially for those man-made slopes constructed for special purposes such as dams and embankments. This method has been widely used in earthquake geotechnical engineering.

In the Rankine Lecture of 1965, Newmark (1965) first proposed the basic elements of a procedure for evaluating the potential displacements of an embankment due to earthquake shaking. Newmark envisaged that sliding would be imminent once the inertia forces on a potential failure block were large enough to overcome the yield resistance and that movement would stop when the inertia forces were reversed.

In his analysis, a soil mass moving downward along a failure surface under inertia force due to earthquake shaking is considered to be analogous to a rigid block with weight and an external force sliding on an inclined plane as shown in Fig. 2.2. In this method, a pseudo-static analysis is first used to calculate a critical or yield acceleration value a_c. The permanent displacement for a sliding mass is then calculated by double integration of the earthquake acceleration time history data above the critical acceleration. Newmark showed that the critical or yield

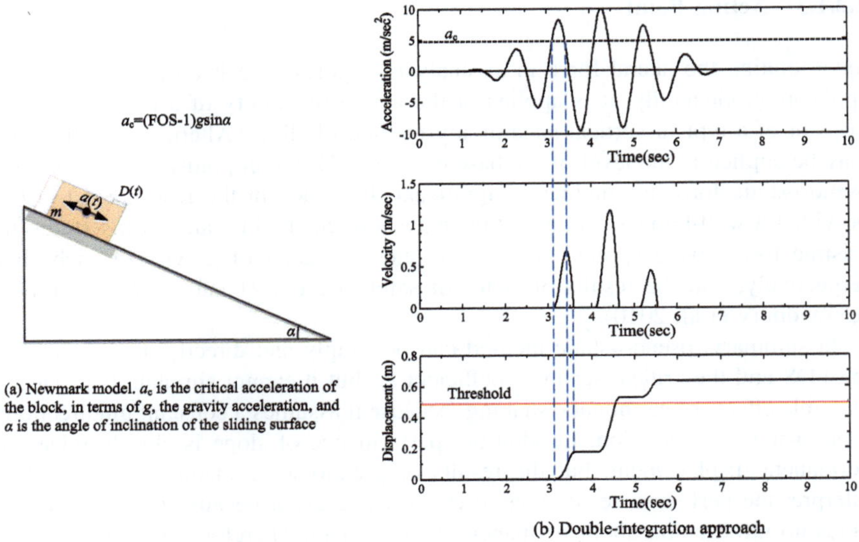

Fig. 2.2 Illustration of the original Newmark's method

acceleration, parallel to the slope, of a potential landslide block is a simple function of the static factor of safety and landslide geometry, expressed as:

$$a_c = (\text{FOS} - 1)g \sin \alpha \tag{2.4}$$

where a_c is in terms of the gravity acceleration g; FOS is the static factor of safety; and α is the angle from the horizontal of the sliding surface.

It should be noted that his concept implied that movements would stop when the inertia forces were reversed. The velocity could remain positive even if the inertia forces were reversed or the inertia forces were not reversed but less than the yield resistance on the potential failure surface. Positive velocity thereby causes sliding on the surface. On the other hand, the velocity could be negative even though the inertia forces were greater than yield resistance. It all depends on the magnitude and direction of both velocity and inertia force, while not either one alone. Besides, as also indicated by Newmark, the uphill resistance without serious error in the calculations may be taken as infinitely large. In this situation, ground motions in the direction of the downward slope tend to move the mass downhill, but ground motions in the upward direction along the slope leave the mass without relative additional motion except where these are extremely large in magnitude. Thus, the negative velocity or velocity heading uphill is not allowed in this analysis.

Since then, the method has been numerous extended and applied. The Sects. 2.2.2.1 and 2.2.2.2 will give reviews for these two aspects, respectively. In addition, a regional scale application of the dynamic sliding block method is reviewed in Sect. 2.2.2.3.

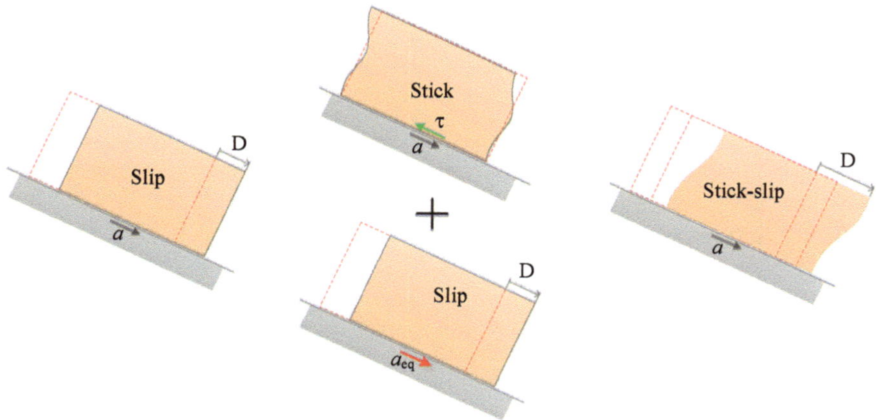

(a) Rigid-block model (b) Decoupled model (c) Coupled model

Fig. 2.3 Calculated models of permanent displacement

2.2.2.1 Extensions

Considerable attention has been focused over the last decades on developing procedures to more accurately analyze the seismic performance of a slope for dams, embankments, or other important structures by modeling the dynamic slope response more rigorously. Soon after Newmark published his rigid-block method, more sophisticated analyses were developed to account for the fact that sliding masses are not rigid bodies but deform internally when subjected to seismic shaking (Seed and Martin 1966; Lin and Whitman 1983). The most commonly used of such analyses was developed by Makdisi and Seed (1977) and calibrated to earth dam using a small number of strong motion records. They produced design charts for estimating seismic displacements as a function of slope geometry, earthquake magnitude, and the ratio of critical or yield acceleration to peak acceleration. Jibson (2011) grouped analytical procedures for estimating permanent coseismic landslide displacements into three types: (1) rigid-block model (Newmark 1965), (2) decoupled model (Makdisi and Seed 1978; Bray and Rathje 1998), and (3) coupled model (Lin and Whitman 1983; Rathje and Bray 2000; Bray and 2007) (Fig. 2.3).

2.2.2.2 Applications

Since the rigid-block method was published in 1965 by Newmark, it has seen numerous applications, four of which are shown in Fig. 2.4. As reviewed by Garini et al. (2011), the applications in recent years include (1) the seismic deformation analysis of earth dams and embankments (Seed and Martin 1966; Ambraseys and Sarma 1967; Sarma 1975, 1981; Franklin and Chang 1977; Makdisi and Seed 1977;

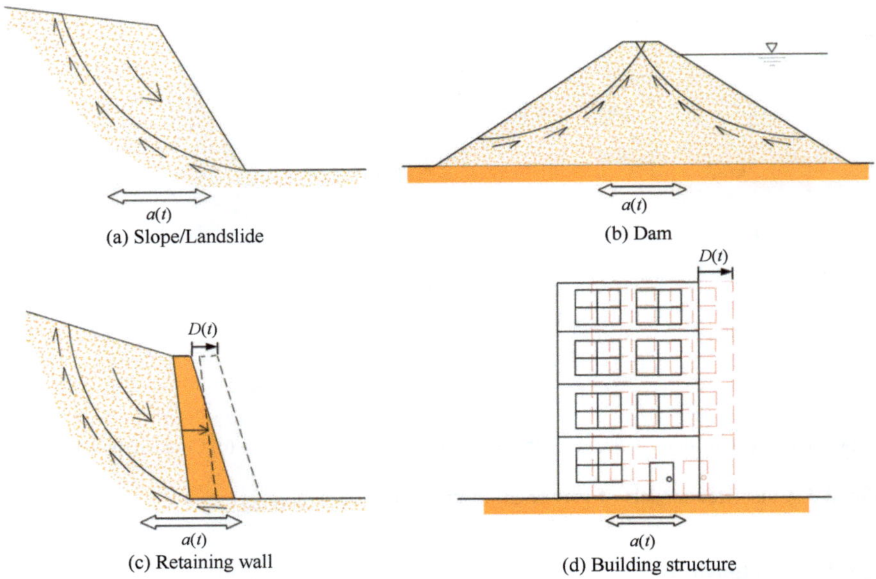

Fig. 2.4 Applications of dynamic sliding block method in geotechnical engineering

Lin and Whitman 1983; Constantinou and Gazetas 1987; Yegian 1991; Yegian et al. 1991; Sawada et al. 1993; Gazetas and Uddin 1994; Kramer 1996; Kramer and Smith 1997; Rathje and Bray 1999); (2) the displacements associated with landslides (Wilson and Keefer 1983; Jibson 1993; Harp and Jibson 1995; Del Gaudio et al. 2003); (3) the seismic deformation of landfills with geosynthetic liners (Bray and Rathje 1998; Yegian et al. 1998); (4) the seismic settlement of surface foundations (Richards et al. 1993); and (5) the potential sliding of concrete gravity dams (Leger and Katsouli 1989; Danay and Adeghe 1993; Fenves and Chopra 1986). The extension of the analogue by Richards and Elms (1979) to gravity retaining walls has met worldwide acceptance and has found its way into seismic codes of practice. Several other generalized applications have also appeared (e.g., Ambraseys and Menu 1988; Ambraseys and Srbulov 1994; Stamatopoulos 1996; Rathje and Bray 2000; Ling 2001; Fardis 2009; Wartman et al. 2003).

2.2.2.3 Regional Scale Analysis

Estimating regions where earthquake-induced landslides are likely to occur and what kind of shaking conditions will trigger them is an important topic in regional landslide seismic hazard assessment (Hsieh and Lee 2011).

Many slope stability analysis methods are not generally applicable for a regional scale analysis. For zonation purposes, it is necessary to quantify in a simple form both the vulnerability of slopes and the triggering mechanisms that can induce instability (Varnes 1984; Rodriguez et al. 1999). With the development of geographic information systems (GIS) tools in recent years, regional scale analyses by the dynamic sliding block method have been proposed. These analyses are based on the integration of seismic shaking parameters, ground geotechnical and geomorphic data including landslide inventories (e.g., Mankelow and Murphy 1998; Miles and Ho 1999; Jibson et al. 1998, 2000; Miles and Keefer 2000; Del Gaudio et al. 2003; Saygili and Rathje 2009). Newmark displacement (Newmark 1965) is commonly taken as a measure of permanent displacement caused by seismic shaking along a slide surface.

The Newmark analysis (which combines slope stability calculations with seismic ground motion records) is widely used to evaluate the potential for landslides that could be triggered by earthquake shaking (Jibson 2000; Jibson and Jibson 2003; Jibson 2007). In essence, Newmark-based seismic landslide hazard or, strictly speaking, spatial variation in slope failure probability or susceptibility, is expressed through regional distribution of Newmark displacements (D_n) that are predicted via empirical relations calibrated using actual seismic landslide records (e.g., Jibson 2000). Subsequent work by Jibson (2007), based on a larger data set of strong motion records (2270 strong motion records from 30 worldwide earthquakes), provided a series of new/updated regression equations to estimate D_n as a function of (i) critical acceleration ratio, (ii) critical acceleration ratio and magnitude, (iii) Arias intensity (I_a) and critical acceleration, and (iv) Arias intensity and critical acceleration ratio. Although all of the regressions have standard deviations of roughly ± 0.5 log units, which correspond to an order of magnitude variation in estimated Newmark displacements, this does not compromise their principal intended use as an indicator of relative hazard that can be quickly obtained for preliminary regional scale assessments (Jibson 2007). Most recently, Hsieh and Lee (2011) used strong motion data from the 1999 Chi-Chi earthquake, the 1999 Kocaeli earthquake, the 1999 Duzce earthquake, the 1995 Kobe earthquake, the 1994 Northridge earthquake, and the 1989 Loma Prieta earthquake to further refine the relationship among critical acceleration (a_c), Arias Intensity (I_a), and Newmark displacement (D_n). After testing different forms of the regression equations, it is shown that an addition of $a_c \log I_a$ term to the Jibson's (1993) empirical formula produces the best results in terms of higher goodness of fit ($R^2 = 0.89$) and lower range of displacements (standard deviation $D_n = 0.295$). Hsieh and Lee (2011) also indicated that it is useful to separate the Taiwanese dataset from the other data and thus developed Taiwan-specific regressions. Similarly, it was shown that the development of separate empirical equations for rock and soil sites leads to statistically better estimates of D_n. Meunier et al. (2007) investigated patterns of well-documented events of widespread seismic landslides in relation to the strong motion data, and their results can been used where accelerometer records and geotechnical information are not available.

2.2.3 Stress-Strain Methods

With the developments of computer technology and simulation approach in recent decades, the numerical simulation method is becoming increasingly used in engineering practice and more and more popular for the real dynamic analysis. These methods can be categorized into continuous methods, e.g., finite element method (FEM) (Clough 1960), finite difference method (FDM) (Mitchell and Griffiths 1980), boundary element method (BEM) (Brebbia and Wrobel 1980), and discontinuous methods, e.g., rigid-block spring method (RBSM) (Kawai 1977, 1978), discrete element method (DEM) (Cundall 1971), and discontinuous deformation analysis (DDA) (Shi and Goodman 1985, 1989).

2.2.3.1 Continuous Methods

Clough (1960) developed and named FEM of engineering analysis based on mathematical methods first developed by Courant (1943) (Jibson 2011). This method uses a mesh to model a deformable system as shown in Fig. 2.5. This method soon began to be applied to slopes, and it provided a valuable tool for modeling the static and dynamic deformation of soil systems.

Several applications of continuous modeling to earth structures have been developed and published; Kramer (1996) provided a good summary of various methods and their associated studies. Seed (1973) analyzed the failures of the Upper and Lower San Fernando dams during the 1971 San Fernando earthquake (M 6.6) by using a finite element model to estimate the strain potential at each node based on cyclic laboratory shear tests of soil samples. Lee (1974) and Serff (1976) used the strain-potential method to model the reduction in the stiffness of soils and thus the permanent slope deformation. More recently, nonlinear inelastic soil models have been developed and implemented in two-dimension (2D) and three-dimension (3D) models (e.g., Prevost 1981; Griffiths and Prevost 1988; Elgamal et al. 1990;

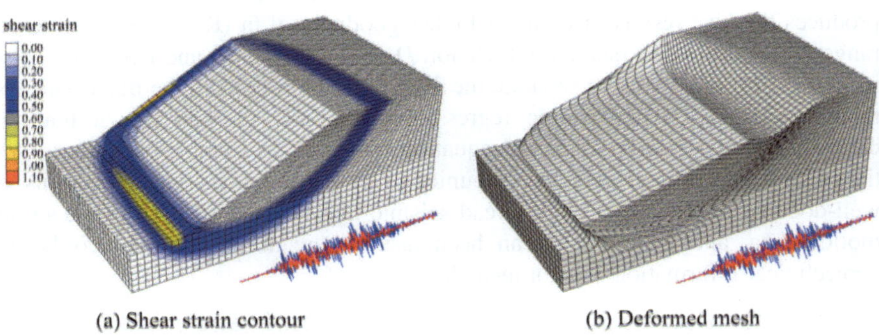

Fig. 2.5 Example of seismic slope analysis by stress-strain methods (Zhang et al. 2013)

Taiebat et al. 2011). In addition, Zheng et al. (2009), Latha and Garaga (2010) and Zhang et al. (2012a) studied the seismic slope stability by using FDM.

2.2.3.2 Discontinuous Methods

For the analysis of a sliding mass consisting of multiple discrete blocks, e.g., a jointed rock slope as shown in Fig. 2.6, the discontinuous methods are more applicable (Ning and Zhao 2012). In the discontinuous methods, the interactions between individual blocks, which will influence the reaction of each block under the seismic loadings, can be conveniently considered. Some applications of RBSM and DEM in such kind of analysis can be found in the literature (Kawai 1977; Kawai et al. 1978, 1981; Niwa et al. 1984; Hamajima et al. 1985) and (Zhang et al. 1997; Papantonopoulos et al. 2002; Psycharis et al. 2003; Bhasin and Kaynia 2004; Pekau and Cui 2004; Komodromos et al. 2008; Papaloizou and Komodromos 2009; Pal et al. 2011), respectively.

DDA is another discontinuous method developed for the modeling of the behaviors of block systems. Since the novel formulation and the numerical code of DDA were presented, DDA draws more and more attention and many modifications and improvements to the original formulas have been proposed to overcome some of its limitations (Ke 1996; Lin et al. 1996; Koo and Chern 1998; Cheng 1998; Doolin and Sitar 2004; Doolin 2005; Wang et al. 2012) and make it more efficient, suitable, and practical to seismic slope stability.

Early applications of DDA to static rock slope engineering can be found in Ohnishi et al. (1995), Ke (1996), Zhao et al. (1997), Wu et al. (1997), Chen and Ohnishi (1999), and Luan et al. (2000). Hatzor et al. (2002, 2004) and Hatzor (2003) demonstrated a fully dynamic 2D DDA stability analysis of a highly discontinuous rock slope using simple rock bolting pattern based on results from dynamic DDA.

In summary, stress-strain method represents a powerful alternative approach for seismic slope stability analysis which is accurate and versatile and requires fewer a

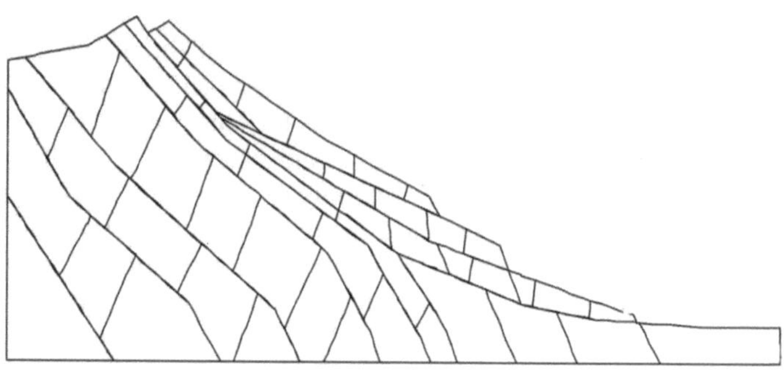

Fig. 2.6 A jointed rock slope (modified from Bhasin and Kaynia 2004)

priori assumptions, especially, regarding the shape of failure surface. However, stress-strain methods use highly complex models that, to be worthwhile, require a high density of high-quality data and sophisticated soil-constitutive models to predict the stress-strain behavior of the soils. For this reason, stress-strain methods are generally practical only for critical projects such as earth dams and slopes affecting critical lifelines or structures. Stress-strain methods are innately site-specific and cannot be applied to regional problems.

2.3 Landslide Run-Out Analysis

The estimation of the movement behavior of a potential landslide is very important for mitigating the landslide disaster. Especially, the run-out distance is one of the major parameters in landslide risk assessment and preventive measure design. Long run-out is one of the major characteristics of earthquake-induced landslides. Many researchers have made great effort to understand how and why large falling masses of rock can move unusually long run-out distance. Researchers have repeatedly revisited the problem using a wide variety of approaches. These efforts can fall into four categories: (1) experimental methods, (2) empirical methods, (3) analytical methods, and (4) numerical simulation methods. The states of the art of these methods are reviewed in the following four Sects. 2.3.1–2.3.4.

2.3.1 Experiment Methods

Physical modelling typically involves using scale models to capture the motion of landslides. Physical experiments are usually preferred to models because models require more assumptions than direct measurements. But for landslides, direct experiment is difficult, dangerous, expensive, and of limited utility. Based on laboratory experiments and filed investigation data, there are many different available models developed for calculating run-out zones.

Some full-scale direct experiments with artificial landslides have been completed (Okura et al. 2000a, b, 2002; Ochiai et al. 2004; Moriwaki et al. 1985; and others). However, since landslides are frequently heterogeneous and single event cannot be repeated carefully through adjusting only one factor, direct experiment is difficult, dangerous, expensive, and of limited utility. And observing conditions are complicated by the danger of being in close proximity to a landslide and the difficulty of measuring a material with properties that change when observed in-situ or when isolated for measurement. But laboratory experiments are still the first qualitative and quantitative observations on the obtained results became fundamental for a better understanding of movement run-out behaviour.

2.3.2 Empirical Methods

Several empirical methods for assessing landslide travel distance and velocity have been developed based on historical data and on the analysis of the relationship between parameters characterizing both the landslide, e.g., the volume of the landslide mass, and the path, e.g., local morphology, and the distance travelled by the failure mass (Hungr and Evans 2004). Regression model-based methods and geomorphology-based methods are two kinds of common methods.

2.3.2.1 Regression Model-Based Methods

The regression model-based methods are developed on an apparent inverse relationship between landslide volume and angle of reach [also called as fahrböschund by Heim (1932)]. Several linear regression equations have been proposed (Scheidegger 1973; Li 1983; Corominas 1996). Introduced by Heim (1932), the angle of reach is the inclination of the line connecting the crest of the source with the toe of the deposit, as measured along the approximate streamline of motion as shown in Fig. 2.7. The angle of reach is considered an index of the efficiency of energy dissipation, and so is inversely related to mobility. Similar correlations between volume and other simple mobility indices have been proposed (Hsü 1975; Davies 1982; Rickenmann 1999). Given estimated source location, volume, and path direction, these methods provide estimates of the distal limit of motion (McDougall 2006a, b).

An alternative principle suggested by Evans and Hungr (1993) is the *minimum shadow angle*, following Lied (1977). This is the angle of a line between the highest point of the talus slope and the stopping point of the longest run-out. According to an investigation of 16 talus slopes in British Columbia by Evans and Hungr (1993), the *minimum shadow angle* is more preferable, but both should only be used for a first approximation of run-out zone.

Keylock and Dommas (1999) tested three empirical models on their ability to predict the maximum length of run-out zone. Their first model was the *height*

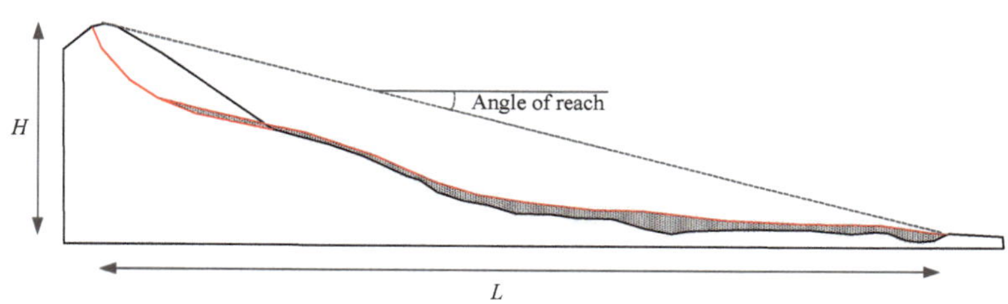

Fig. 2.7 Schematic definition of the angle of reach. H is the elevation difference between the crest of the source and the toe of the deposit, and L is the length of the horizontal projection of the streamline connecting these two points

function model, in which the run-out distance derived from the combined vertical height of the free rock face and the talus slope. The second was the *α-β model* following Heim (1932), Hsü (1975), and Köner (1980), which is based on the correlation between the average energy of an extreme rockfall event. Their third model was *run-out ratio model*, developed by McClung and Lied (1987), which describes the ratio between the horizontal length of the run-out zone to the combined horizontal length of the talus slope and the free rock face. After investigated using rockfall data presented by Domaas (1994), the most accurate model of the three appeared to the *run-out ratio model.*

Improved empirical model notable performing regressions on subsets with varying scopes were presented by Cannon (1993), Corominas (1996), Jakob and Hungr (2005), and others.

Regression model-based models play a valuable role in landslide run-out analysis due to the regression model-based methods are simple. But the regression model-based methods are difficult to apply in practice with a high degree of certainty. For example, the correlation coefficients for some of regression models are 0.7–0.8, while a value of larger than 0.95 generally indicates a strong correlation. And it is difficult in this method to take account of influences of the ground condition, the micro-topography, the degree of saturation of the landslide mass. For this point, geomorphology-based method is another alternative approach to predict the run-out of landslide.

2.3.2.2 Geomorphology-Based Methods

Field work and photograph interpretation are the main sources of the geomorphological analysis for determining the travel distance of landslides (Hungr et al. 2004). The assessment of the extent of both ancient and recent landslide deposits is the basis for defining future travel distances. The outer margin of the landslide deposits gives an appraisal of the maximum distances that landslides have been able to reach during the present landscape (Fig. 2.8). Several authors have provided these studies (e.g., Costa 1984; Kostashuk 1987; Copons et al. 2004; Copons and Vilaplana 2008).

The geomorphological approach does not give any clue of the emplacement mechanism. Furthermore, the slope geometry and the circumstances responsible for past landslides might have changed. Therefore, results obtained in a given place cannot be easily exported to other localities.

In summary, empirical methods, both regression model-based methods and geomorphology-based methods, typically predict travel distances, while the deformation characteristics or the slide velocities of the landslide are not predicted. These models may be applied to establish initial hazard characteristics for preliminary run-out analysis, which may be later refined by other models.

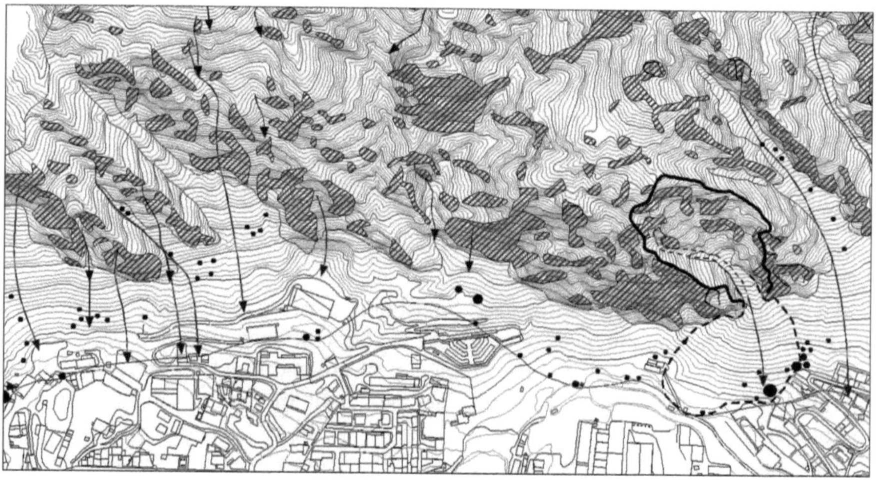

Fig. 2.8 Boundary of the potential rockfall run-out area in Santa Coloma (Principality of Andorra), defined by the line that links the farthest fallen blocks observed in the field (Copons 2004). *Arrows* indicate historical rockfall paths and solid circles are large fallen boulders (Hungr et al. 2004)

2.3.3 Analytical Methods

In contrast to empirical methods, analytical methods are based on mechanics and involve the solution of motion equations (McDougall 2006a, b). The simplest analytical model is the classical sliding block model as shown in Fig. 2.9, which is based on work-energy theory (Kirby and Statham 1975; Bozzolo and Pamini 1986; Hungr and Evans 1988; Pfeiffer and Bowen 1989; Kobayashi et al. 1990; Evans and Hungr 1993; Azzoni et al. 1995; Chau et al. 1996; Bozzolo and Pamini 1986; Azzoni et al. 1995; Müller-Bernet in Heim 1932; Sassa 1988). Internal deformation and its associated energy dissipation are neglected and the landslides is treated as a lumped mass. At any position along the path, the sum of the energies including the potential energy, kinetic energy, and net energy loss equals the initial potential energy. This energy balance can be visualized using the concept of energy grade lines, as shown in Fig. 2.9. The concept of energy grade lines is useful for visualizing the energy balance. v is the velocity of the block, g is the vertical acceleration due to gravity, and $v^2/2\ g$ is known as the velocity head, which is the kinetic energy of the block normalized by the product of its mass and g. The same normalization of net energy loss is known as head loss. Note that the positions of the energy lines are referenced to the centre of mass of the block and that the true energy line and mean energy line do not necessarily coincide. Given the initial position of the center of mass and a suitable relationship to approximate the energy losses, the position and velocity of the block can be determined at any given time.

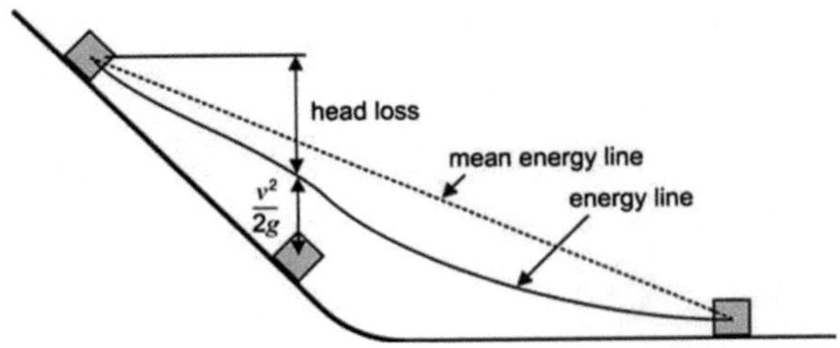

Fig. 2.9 Classical sliding block model, based on work-energy theory (McDougall 2006a, b)

To compute the velocity before and after bounce, two principle approaches were used both considering the basis of energy loss after a bounce. One approach defined energy loss by a coefficient for the efficiency of collision and the other calculated energy loss on the basis of a tangential coefficient of restitution and a normal coefficient.

Three dimensional for investigating run-out at a slope scale were also developed (Descoeudres and Zimmermann 1987; Guzzetti el al. 2002; Dorren et al. 2004; Lan et al. 2007 and Masuya et al. 2009). These models require a high-resolution digital elevation model (DEM).

Generally speaking, the use of analytical methods is somewhat motivated by the limitations of purely empirical methods, as the unique geometry and materials involved in each case can be accounted for explicitly and a statistically significant database of previous events is not necessarily required. The simplicity of a lumped mass allows analytical solutions, fast and effectively (Hürlimann et al. 2008). However, because the landslide is reduced to a single point, lumped mass models cannot provide the exact maximum run-out distance, but only the displacement concerning the centre of mass (Evans et al. 1994; Hungr 1995).

2.3.4 *Numerical Simulation Methods*

The single-block model should be only applied to the motion of the center of mass of a rigid body, but more complex continuum deformable mass or multi-block system is often appeared in practice. Some numerical simulation methods have been developed to account explicitly for deformation during motion.

2.3.4.1 Continuous Methods

When considering that the dimensions of a typical particle is much smaller than the depth and length of the debris, the debris mass is treated as continuum. According to depth-averaged Saint Venant approach, the material is assumed to be incompressible and the mass and momentum equations are written in a depth-averaged form. Many numerical methods now exist to investigate the run-out process of landslide (e.g., Savage and Hutter 1989; Sousa and Voight 1991; Takahashi et al. 1992; Hungr 1995; Chen and Lee 2000; Denlinger and Iverson 2001; Crosta et al. 2003, 2005, 2007; McDougall and Hungr 2004; Pirulli 2005; McDougall 2006a, b). These methods are usually based on continuum mechanics and assume that the avalanche thickness is very much smaller than its extent parallel to the bed, i.e., thin layer depth-averaged models. The primary differences are their representation of basal resistance force and the constitutive relations describing the mechanical behavior of the considered material. These models can accurately take account of detailed topography effects, shown to be significant, with a reasonable computational time, making it possible to perform sensitivity studies of the parameters used in the model. They can provide effective properties that make it possible to roughly reproduce not only the deposit shape, but also the dynamic as shown in Favreau et al. (2010) and Moretti et al. (2012) for examples. However, conventional continuum approaching models, which neglects the contact between rocks, makes it impossible to trace the position of individual rock during a landslide.

2.3.4.2 Discontinuous Methods

When the landslide mass consists of large fragments and boulders, the run-out mass is modelled as an assembly of blocks moving down a surface. By applying known individual constitutive properties, contact laws, velocities, displacements, and body forces, their dynamic behavior can be studied over a selected period in time. Some authors take circular shape models in their run-out analysis to evaluate maximum run-out and final deposit position of past or potential events (e.g., Poisel et al. 2008). Although polygonal shapes have the disadvantages due to the complexity of the contact patterns and penalty in computational time, methods using non-circular shapes will be required for more real-world problems. It is more appropriate when problems are limited in finite blocks. Discontinuous numerical simulation methods are powerful tools in simulation of failure and run-out process of rock avalanche controlled by weakness surface. DEM (Cundall 1971) and DDA (Shi and Goodman 1985, 1989; Shi 1988) are two of the most commonly used methods.

Both DEM and DDA employ the equations of dynamic motion which are solved at finite points in time, in a series of time steps, but there are some subtle but significant differences in their formulations of the solution schemes and contact mechanics. In the solution schemes, equations of motion in DDA are derived using the principle of minimization of the total potential energy of the system, while the equations of motion as implemented in DEM are derived directly from the force balance equations, which still resultant unbalanced force after a time step and damping is necessarily used to dissipate energy. In the contact mechanics, the DDA used a penalty method in which the contact is assumed to be rigid. No overlapping or interpenetration of the blocks is allowed as the same as real physical cases, whereas soft contact approach is used in DEM. The soft contact approach requires laboratory or field measured joint stiffness, which may be difficult to obtain in many cases. Many comparisons of basic models (sliding, colliding, and rolling models) between the DEM and DDA were carried out and show that the results from DDA are more close to the analytical values than that from DEM (Zheng 2010). Compared to DEM, DDA has a simpler and more straightforward physical meaning (Wu 2003).

Applications of DEM can be found in some literatures, such as Zhang et al. 1997, Papantonopoulos et al. 2002, Psycharis et al. 2003, Pekau and Cui 2004, Komodromos et al. 2008, Papaloizou and Komodromos 2009.

DDA is a dynamic numerical analysis method capable of evaluating the impacts area of an earthquake-induced landslide when seismic impacts are integrated into simulations. Hatzor and Feintuch (2001) was the first to validate the use of DDA in simulating dynamic landslides by studying the dynamic of block sliding on an inclined plane, in which they assumed that the base block was fixed and earthquake accelerations were directly considered as body force and added to the sliding block in DDA. Based on the same input model of seismic loadings, Makris and Roussos (2000), Shi (2002), Kong and Liu (2002), Ishikawa et al. (2002), Hatzor et al. (2004), Tsesarsky et al. (2005), Yagoda and Hatzor (2010), and Bakun-Mazor et al. (2012) studied the dynamic response or/and stability analysis of tunnel, slope, dam, foundation, or ancient masonry structure by using DDA. Alternatively, Sasaki et al. (2004) developed an acceleration input method different from the original DDA algorithm to simulate the dynamic behavior of a slope with sliding block. In his method, the seismic accelerations were applied to the base block, which is different from the former seismic loadings input model. Sasaki et al. (2007) applied the same earthquake input model to analysis several cases of simple block structures under harmonic accelerations to acquire the relationships between natural frequencies of elastic block structures and applied accelerations. Later, Wu group (2009, 2010, 2011) applied DDA to simulate the kinematic behavior of sliding rock blocks in the Tsaoling landslide and the Chiu-fen-erh-shan landslide induced by the 1999 Chi-Chi earthquake. Recently, Zhang et al. (2012b) applied newest DDA program to simulate the largest landslides induced by the 2008 Wenchuan earthquake.

2.4 Comparisons of Various Methods and Conclusions

The studies in the field of the earthquake-induced landslides are generally reviewed. Two parts of contents (i) seismic stability analysis and (ii) run-out analysis are reviewed and compared. Some conclusions can be drawn:

(1) Three categories of methods can be used to analyze the seismic stability of a slope. Each of these types of methods has strengths and weaknesses and each can be appropriately applied in different situations. In detail, pseudo-static methods can simply and directly determine the FOS and the critical coefficient k_c of a slope, while the widely used Newmark's methods and its extensions can

determine the coseismic deformation of a slope. And the Newmark's methods can be used to estimate where earthquake-induced landslides are likely to occur and what kind of shaking conditions will trigger them based on the GIS technology. More sophisticated analysis for real dynamic process of a seismic slope should be carried out by stress-strain methods, including both continuous methods and discontinuous methods.

(2) Four kinds of methods can be used to analyze the run-out of a landslide. In detail, experiment method can provide the qualitative and quantitative observations on the obtained results although this method is difficult, dangerous, expensive, and of limited utility. Empirical method can be directly used for assessing landslide travel distance and velocity based on historical data and on the analysis of the relationship between parameters characterizing both landslide and the path. Analytical method can be more directly used without the need of statistically significant database of previous events. Numerical simulation method can be used to provide more information for the landslide composed by the complex continuum deformable mass or multi-blocks.

References

Ambraseys, N., & Sarma, S. (1967). The response of earth dams to strong earthquakes. *Geotechnique, 17*(3), 181–213.

Ambraseys, N., & Menu, J. (1988). Earthquake-induced ground displacements. *Earthquake Engineering & Structural Dynamics, 16*(7), 985–1006.

Ambraseys, N., & Srbulov, M. (1994). Attenuation of earthquake-induced ground displacements. *Earthquake Engineering & Structural Dynamics, 23*(5), 467–487.

Anderson, D., & Kavazanjian Jr, E. (1995). Performance of landfills under seismic loading. In *Proc., 3rd Int. Conf. on Recent Advances in Geotechnical Earthquake Engineering and Soil Dynamics*. (Vol.3, pp. 277–306) Rolla, MO: Univ. of Missouri.

Azzoni, A., La Barbera, G., & Zaninetti, A. (1995). Analysis and prediction of rockfalls using a mathematical model. In International Journal of Rock Mechanics and Mining Sciences & Geomechanics Abstracts. *Elsevier, 32,* 709–724.

Bakun-Mazor, D., Hatzor, Y., & Glaser, S. (2012). Dynamic sliding of tetrahedral wedge: the role of interface friction. *International Journal for Numerical and Analytical Methods in Geomechanics, 36*(3), 327–343.

Bhasin, R., & Kaynia, A. M. (2004). Static and dynamic simulation of a 700-m high rock slope in western Norway. *Engineering Geology, 71*(3–4), 213–226.

Bozzolo, D., & Pamini, R. (1986). Simulation of rock falls down a valley side. *Acta Mechanica, 63* (1–4), 113–130.

Bray, J. D., & Rathje, E. M. (1998). Earthquake-induced displacements of solid-waste landfills. *Journal of Geotechnical and Geoenvironmental Engineering, 124*(3), 242–253.

Bray, J. D., & Travasarou, T. (2007). Simplified procedure for estimating earthquake-induced deviatoric slope displacements. *Journal of Geotechnical and Geoenvironmental Engineering, 133*(4), 381–392.

Brebbia, C.A., & Wrobel, L. (1980). The boundary element method. *Computer methods in fluids.* (A 81-28303 11-34) (pp. 26–48). London: Pentech Press, Ltd..

Cannon, S.H. (1993). An empirical model for the volume-change behavior of debris flows. In *Hydraulic Engineering '93*. San Francisco.

CDCDMG: California Department of Conversation, Division of Mines and Geology. (1997). *Guidelines for evaluating and mitigating seismic hazards in California*. CDMG Special Publication.

Chau, K., Wong, R., & Lee, C. (1996). Rockfall problems in Hong Kong and some new experimental results for coefficients of restitution. In International Journal of Rock Mechanics and Mining Sciences and Geomechanics Abstracts. *Elsevier, 35,* 662–663.

Chen, G., & Ohnishi, Y. (1999). Slope stability analysis using discontinuous deformation Analysis method. *Rock Mecganics for Industry* (pp. 535–541).

Chen, H., & Lee, C. (2000). Numerical simulation of debris flows. *Canadian Geotechnical Journal, 37*(1), 146–160.

Cheng, Y. M. (1998). Advancements and improvement in discontinuous deformation analysis. *Computers and Geotechnics, 22*(2), 153–163.

Chopra, A. K. (1966). The importance of the vertical component of earthquake motions. *Bulletin of the Seismological Society of America, 56*(5), 1163–1175.

Chowdhury, R., Flentje, P., & Bhattacharya, G. (2010). *Geotechnical slope analysis*. CRC PressI Llc.

Clough, R.W. (1960). The finite element method in plane stress analysis. In *2nd Conference on Electronic Computation*. Pittsburgh, PA.

Constantinou, M., & Gazetas, G. (1987). Probabilistic seismic sliding deformations of earth dams and slopes. In *Probabilistic Mechanics and Structural Reliability (1984)* (pp. 318–321). ASCE.

Copons, R., & Vilaplana, J. M. (2008). Rockfall susceptibility zoning at a large scale: From geomorphological inventory to preliminary land use planning. *Engineering Geology, 102*(3), 142–151.

Copons, R., Vilaplana, J.M., Corominas, J., Altimir, J., & Amigó, J. (2004). Rockfall risk management in high-density urban areas. The andorran experience. *Landslide Hazard and Risk*, 675–698.

Corominas, J. (1996). The angle of reach as a mobility index for small and large landslides. *Canadian Geotechnical Journal, 33*(2), 260–271.

Costa, J.E. (1984). Physical geomorphology of debris flows. In *Developments and applications of geomorphology* (pp. 268–317). Springer.

Courant, R. (1943). Variational methods for the solution of problems of equilibrium and vibrations. *Bulletin American Mathematics Society, 49*(1), 23.

Crosta, G., & Agliardi, F. (2003). A methodology for physically based rockfall hazard assessment. *Natural Hazards and Earth System Science, 3*(5), 407–422.

Crosta, G., & Frattini, P. (2003). Distributed modelling of shallow landslides triggered by intense rainfall. *Natural Hazards and Earth System Science, 3*(1/2), 81–93.

Crosta, G.B., Frattini, P., & Fusi, N. (2007). Fragmentation in the Val Pola rock avalanche, italian alps. *Journal of Geophysical Research: Earth Surface* (2003–2012), 112(F1).

Crosta, G., Imposimato, S., Roddeman, D., Chiesa, S., & Moia, F. (2005). Small fast-moving flow-like landslides in volcanic deposits: the 2001 Las Colinas Landslide (El Salvador). *Engineering Geology, 79*(3), 185–214.

Cundall, P. (1971). A computer model for simulating progressive, large scale movements in blocky rock system. In *Symposium of International Society of Rock Mechanics* (pp. 11–18). France: Nancy.

Danay, A., & Adeghe, L. (1993). Seismic-induced slip of concrete gravity dams. *Journal of Structural Engineering, 119*(1), 108–129.

Davies, T. R. (1982). Spreading of rock avalanche debris by mechanical fluidization. *Rock Mechanics, 15*(1), 9–24.

Del Gaudio, V., Pierri, P., & Wasowski, J. (2003). An approach to time-probabilistic evaluation of seismically induced landslide hazard. *Bulletin of the Seismological Society of America, 93*(2), 557–569.

Denlinger, R.P., & Iverson, R.M. (2001). Flow of variably fluidized granular masses across three-dimensional terrain: 2. Numerical predictions and experimental tests. *Journal of Geophysical Research: Solid Earth* (1978–2012), *106*(B1): 553–566.

Descoeudres, F., & Zimmermann, T. (1987). Three-dimensional dynamic calculation of rockfalls. In *6th ISRM Congress.*

Domaas, U. (1994). *Geometrical methods of calculating rockfall range* (NGI Report: 585910–585911).

Doolin, D. M. (2005). Unified displacement boundary constraint formulation for discontinuous deformation analysis (DDA). *International Journal for Numerical and Analytical Methods in Geomechanics, 29*(12), 1199–1207.

Doolin, D. M., & Sitar, N. (2004). Time integration in discontinuous deformation analysis. *Journal of Engineering Mechanics, 130*(3), 249–258.

Dorren, L., & Heuvelink, G. B. (2004). Effect of support size on the accuracy of a distributed rockfall model. *International Journal of Geographical Information Science, 18*(6), 595–609.

Elgamal, A.-W. M., Scott, R. F., Succarieh, M. F., & Yan, L. (1990). La Villita dam response during five earthquakes including permanent deformation. *Journal of Geotechnical Engineering, 116*(10), 1443–1462.

Evans, S., & Hungr, O. (1993). The assessment of rockfall hazard at the base of talus slopes. *Canadian Geotechnical Journal, 30*(4), 620–636.

Evans, S., Hungr, O., & Enegren, E. (1994). The Avalanche Lake rock avalanche, Mackenzie mountains, northwest territories, Canada: description, dating, and dynamics. *Canadian Geotechnical Journal, 31*(5), 749–768.

Fardis, M.N. (2009). Seismic design, assessment and retrofitting of concrete buildings: based on EN-Eurocode 8. Springer.

Favreau, P., Mangeney, A., Lucas, A., Crosta, G., & Bouchut, F. (2010). Numerical modeling of landquakes. *Geophysical Research Letters, 37*(15), L15305.

Fenves, G., & Chopra, A.K. (1986). Simplified analysis for earthquake resistant design of concrete gravity dams. Earthquake Engineering Research Center, University of California.

Franklin, A.G., & Chang, F.K. (1977). Permanent displacements of earth embankments by Newmark sliding block analysis.

Garini, E., Gazetas, G., & Anastasopoulos, I. (2011). Asymmetric 'Newmark' sliding caused by motions containing severe 'directivity' and 'fling' pulses. *Geotechnique 61*, 733–756.

Gazetas, G., & Uddin, N. (1994). Permanent deformation on preexisting sliding surfaces in dams. *Journal of Geotechnical Engineering, 120*(11), 2041–2061.

Griffiths, D., & Prevost, J. H. (1988). Two-and three-dimensional dynamic finite element analyses of the Long Valley Dam. *Geotechnique, 38*(3), 367–388.

Guzzetti, F., Malamud, B. D., Turcotte, D. L., & Reichenbach, P. (2002). Power-law correlations of landslide areas in central Italy. *Earth and Planetary Science Letters, 195*(3), 169–183.

Hamajima, R., Kawai, T., Yamashita, K., & Kusabuka, M. (1985). Numerical analysis of cracked and jointed rock mass. In *the 5th International Conference on Numerical Methods in Geomechanics* (pp. 207–214). Nagoya, Japan.

Harp, E. L., & Jibson, R. W. (1995). *Inventory of landslides triggered by the 1994 Northridge*. US Geological Survey: California earthquake.

Hatzor, Y. H. (2003). Fully dynamic stability analysis of jointed rock slopes. In *Proceedings of the 10th ISRM Congress*, (pp. 503–514).

Hatzor, Y. H., & Feintuch, A. (2001). The validity of dynamic block displacement prediction using DDA. *International Journal of Rock Mechanics & Mining Sciences.*

Hatzor, Y. H., Arzi, A. A., & Tsesarsky, M. (2002). Realistic dynamic analysis of jointed rock slopes using DDA. In *5th Int. Conf. on Analysis of Discontinuous Deformation—Stability of rock structures*, Abingdon (pp. 47–56). Rotterdam, The Netherlands: Balkema.

Hatzor, Y., Arzi, A. A., Zaslavsky, Y., & Shapira, A. (2004). Dynamic stability analysis of jointed rock slopes using the DDA method: King Herod's Palace, Masada, Israel. *International Journal of Rock Mechanics and Mining Sciences, 41*(5), 813–832.

Heim, A. (1932). *Bergsturz und Menschenleben*. Zurich: Fretz and Wasmuth Verlag.

Hsieh, S.-Y., & Lee, C.-T. (2011). Empirical estimation of the Newmark displacement from the Arias intensity and critical acceleration. *Engineering Geology, 122*(1–2), 34–42.

Hsü, K. J. (1975). Catastrophic debris streams (sturzstroms) generated by rockfalls. *Geological Society of America Bulletin, 86*(1), 129–140.

Hungr, O. (1995). A model for the runout analysis of rapid flow slides, debris flows, and avalanches. *Canadian Geotechnical Journal, 32*(4), 610–623.

Hungr, O., & Evans, S. (1988). Engineering evaluation of fragmental rockfall hazards. In *Proceedings of the Fifth International Symposium on Landslides, Lausanne* (pp. 685–690). Rotterdam, Netherlands: AA Balkema.

Hungr, O., & Evans, S. (2004). Entrainment of debris in rock avalanches: an analysis of a long run-out mechanism. *Geological Society of America Bulletin, 116*(9–10), 1240–1252.

Hungr, O., Corominas, J., & Eberhardt, E. (2004). Estimating landslide motion mechanism, travel distance and velocity.

Hürlimann, M., Rickenmann, D., Medina, V., & Bateman, A. (2008). Evaluation of approaches to calculate debris-flow parameters for hazard assessment. *Engineering Geology, 102*(3), 152–163.

Hynes-Griffin, M. E., & Franklin, A. G. (1984). Rationalizing the seismic coefficient method. Defense Technical Information Center.

Ishikawa, T., Sekine, E., & Ohnishi, Y. (2002). Shaking table tests of coarse granular materials with discontinuous analysis. In *Proc. of ICADD-5, BALKEMA*, (pp. 181–187).

Jakob, M., & Hungr, O. (2005). Debris-flow hazards and related phenomena. Springer.

Jibson, R.W. (1993). Predicting earthquake-induced landslide displacements using Newmark's sliding block analysis. *Transportation Research Record*, 9–9.

Jibson, R. (2000). A method for producing digital probabilistic seismic landslide hazard maps. *Engineering Geology.*

Jibson, R. W. (2007). Regression models for estimating coseismic landslide displacement. *Engineering Geology, 91*(2–4), 209–218.

Jibson, R. W. (2011). Methods for assessing the stability of slopes during earthquakes—a retrospective. *Engineering Geology, 122*(1–2), 43–50.

Jibson, R. W., & Jibson, M. W. (2003). *Java programs for using Newmark's method and simplified decoupled analysis to model slope performance during earthquakes.* US Geological Survey: US Department of the Interior.

Jibson, R. W., Harp, E. L., & Michael, J. A. (1998). A Method for Producing Digital Probabilistic Seismic Landslide Hazard Maps: An Example from the Los Angeles, California, Area.

Kavazanjian, E., & Consultants, G. (1997). *Design Guidance: Geotechnical Earthquake Engineering for Highways.* Federal Highway Administration: Design Principles.

Kawai, T. (1977). A new discrete analysis of nonlinear solid mechanics problems involving stability, plasticity and crack. *In the Symposium on Applications of Computer Methods in Engineering, Los Angeles, USA*, (pp. 1029–1038).

Kawai, T. (1978). New discrete models and their application to seismic response analysis of structures. *Nuclear Engineering and Design, 48*(1), 207–229.

Kawai, T., Kawabata, Y., Kumagai, K., & Kondou, K. (1978). A new discrete model for analysis of solid mechanics problems. *Numerical methods in fracture mechanics*, 26–37.

Kawai, T., Takeuchi, N., & Kumeta, T. (1981). New discrete models and their application to rock mechanics. In *ISRM International Symposium.*

Ke, T. C. (1996). The issues of rigid-body rotation in DDA. In *First international forum on discontinuous deformation analysis (DDA) and simulations of discontinuous media,* (pp. 318–325). Berkeley, USA.

Keefer, D. K. (2000). Statistical analysis of an earthquake-induced landslide distribution—the 1989 Loma Prieta California event. *Engineering Geology, 58*(3), 231–249.

Keylock, C., & Domaas, U. (1999). Evaluation of topographic models of rockfall travel distance for use in hazard applications. *Arctic, Antarctic, and Alpine Research*, 312–320.

Kirby, M. J., & Statham, I. (1975). Surface stone movement and screen formation. *Journal of Geology, 83*(3), 349–362.

Kobayashi, Y., Harp, E., & Kagawa, T. (1990). Simulation of rockfalls triggered by earthquakes. *Rock Mechanics and Rock Engineering, 23*(1), 1–20.

Komodromos, P., Papaloizou, L., & Polycarpou, P. (2008). Simulation of the response of ancient columns under harmonic and earthquake excitations. *Engineering Structures, 30*(8), 2154–2164.

Kong, X., & Liu, J. (2002). Dynamic failure numeric simulations of model concrete-faced rock-fill dam. *Soil Dynamics and Earthquake Engineering, 22*(9–12), 1131–1134.

Koo, C. Y., & Chern, J. C. (1998). Modification of the DDA method for rigid block problems. *International Journal of Rock Mechanics & Mining Sciences, 35,* 683–693.

Körner, H. (1980). The energy-line method in the mechanics of avalanches. *Journal of Glaciology, 26,* 501–505.

Kostaschuk, R. (1987). Identification of debris flow hazard on alluvial fans in the Canadian Rocky Mountains. *Debris flows/avalanches: process, recognition, and mitigation, 7,* 115.

Kramer, S.L. (1996). *Geotechnical earthquake engineering*. Prentice-Hall Civil Engineering and Engineering Mechanics Series, Upper Saddle River, NJ: Prentice Hall,| c1996, 1.

Kramer, S. L., & Smith, M. W. (1997). Modified Newmark model for seismic displacements of compliant slopes. *Journal of Geotechnical and Geoenvironmental Engineering, 123*(7), 635–644.

Krinitzsky, E.L. (1993). Fundamentals of earthquake-resistant construction. Wiley. com.

Lan, H., Derek Martin, C., & Lim, C. (2007). RockFall analyst: a GIS extension for three-dimensional and spatially distributed rockfall hazard modeling. *Computers & Geosciences, 33*(2), 262–279.

Latha, G. M., & Garaga, A. (2010). Seismic stability analysis of a himalayan rock slope. *Rock Mechanics and Rock Engineering, 43*(6), 831–843.

Lee, K. L. (1974). *Seismic permanent deformation in earth dams*. Los Angeles, CA: University of California.

Leger, P., & Katsouli, M. (1989). Seismic stability of concrete gravity dams. *Earthquake Engineering & Structural Dynamics, 18*(6), 889–902.

Li, T. (1983). A mathematical model for predicting the extent of a major rockfall. *Zeitschrift Fur Geomorphologie, 24,* 473–482.

Lied, K. (1977). Rockfall problems in Norway. Rockfall dynamics and protective work effectiveness. ISMES publ, 90, 51–53.

Lin, J. S., & Whitman, R. V. (1983). Decoupling approximation to the evaluation of earthquake-induced plastic slip in earth dams. *Earthquake Engineering & Structural Dynamics, 11*(5), 667–678.

Lin, C.T., Amadei, B., Jung, J., & Dwyer, J. (1996). Extensions of discontinuous deformation analysis for jointed rock masses. *international Journal of Rock Mechanics and Mining Sciences & Geomechanics Abstracts*, *33*(7), 671–694.

Ling, H. I. (2001). Recent applications of sliding block theory to geotechnical design. *Soil Dynamics and Earthquake Engineering, 21,* 189–197.

Ling, H., & Leshchinsky, D. (1998). Effects of vertical acceleration on seismic design of geosynthetic-reinforced soil structures. *Geotechnique, 48*(3), 347–373.

Luan, M., Li, Y., & Yang, Q. (2000). Discontinuous deformation computational mechanics model and its application in stability analysis of rock slope. *Chinese Journal of Rock Mechanics and Engineering, 3,* 006.

Makdisi, F.I., & Seed, H.B. (1977). Simplified procedure for estimating dam and embankment earthquake-induced deformations. In *ASAE Publication No. 4-77. Proceedings of the National Symposium on Soil Erosion and Sediment by Water*, Chicago, Illinois, December 12–13.

Makdisi, F. I., & Seed, H. B. (1978). Simplified procedure for estimating dam and embankment earthquake-induced failures. *Journal of the Geotechnical Division, ASCE, 104,* 849–861.

Makris, N., & Roussos, Y. (2000). Rocking response of rigid blocks under near-source ground motions. *Geotechnique, 50*(3), 243–262.

Mankelow, J.M., & MURPHY, W. (1998). Using GIS in the probabilistic assessment of earthquake triggered landslide hazards. *Journal of Earthquake Engineering*, *2*(4), 593–623.

Marcuson, W. (1981). Moderator's report for session on Earth Dams and Stability of Slopes under Dynamic Loads. In *Proceedings, International Conference on Recent Advances in Geotechnical Earthquake Engineering and Soil Dynamics* (Vol.3, p. 1175).

Marcuson III, W.F., & Franklin, A.G. (1983). *Seismic Design, Analysis, and Remedial Measures to Improve Stability of Existing Earth Dams*. DTIC Document.

Masuya, H., Amanuma, K., Nishikawa, Y., & Tsuji, T. (2009). Basic rockfall simulation with consideration of vegetation and application to protection measure. *Natural Hazards and Earth System Science, 9*(6), 1835–1843.

McClung, D. M., & Lied, K. (1987). Statistical and geometrical definition of snow avalanche runout. *Cold Regions Science and Technology, 13*(2), 107–119.

McDougall, S. (2006a). *A new continuum dynamic model for the analysis of extremely rapid landslide motion across complex 3D terrain*. University of British Columbia.

McDougall, D. (2006b). The distributed criterion design. *Journal of Behavioral Education, 15*(4), 236–246.

McDougall, S., & Hungr, O. (2004). A model for the analysis of rapid landslide motion across three-dimensional terrain. *Canadian Geotechnical Journal, 41*(6), 1084–1097.

Meunier, P., Hovius, N., & Haines, A.J. (2007). Regional patterns of earthquake-triggered landslides and their relation to ground motion. *Geophysical Research Letters, 34*(20).

Miles, S. B., & Ho, C. L. (1999). Applications and issues of GIS as tool for civil engineering modeling. *Journal of Computing in Civil Engineering, 13*(3), 144–152.

Miles, S. B., & Keefer, D. K. (2000). Evaluation of seismic slope-performance models using a regional case study. *Environmental & Engineering Geoscience, 6*(1), 25–39.

Mitchell, A.R., & Griffiths, D.F. (1980). *The finite difference method in partial differential equations*. Wiley-Interscience: Chichester, Sussex, England and New York. p. 281.

Moretti, L., Mangeney, A., Capdeville, Y., Stutzmann, E., Christian Huggel, C., Schneider, D., & Francois Bouchut, F. (2012). Numerical modeling of the mount steller landslide flow history and of the generated long period seismic waves. *Geophysical Research Letters, 39*(L16402).

Moriwaki, H., Yazaki, S., & Oyagi, N. (1985). A gigantic debris avalanche and its dynamics at Mount Ontake caused by the Nagano-ken-seibu earthquake, 1984. In *Proc. 4th Int. Conf. Field Workshop on Landslides*, (pp. 359–364).

Newmark, N. M. (1965). Effects of earthquakes on dams and embankments. *Géotechnique, 15*, 139–159.

Ning, Y., & Zhao, Z. (2012). A detailed investigation of block dynamic sliding by the discontinuous deformation analysis. *International Journal for Numerical and Analytical Methods in Geomechanics*, 1–21.

Niwa, K., Kawai, T., Ikeda, M., & Takeda, T. (1984). Application of a new discrete method to fracture analysis of brittle materials. In *The 3rd International Conference on Numerical Methods in Fracture Mechanics, Swansea, U.K.* (pp. 13–27).

Ochiai, H., Okada, Y., Furuya, G., Okura, Y., Matsui, T., Sammori, T., et al. (2004). A fluidized landslide on a natural slope by artificial rainfall. *Landslides, 1*(3), 211–219.

Ohnishi, Y., Chen, G., & Miki, S. (1995). Recent development of DDA in rock mechanics. *Proc. ICADD, 1*, 26–47.

Okura, Y., Kitahara, H., & Sammori, T. (2000a). Fluidization in dry landslides. *Engineering Geology, 56*(3), 347–360.

Okura, Y., Kitahara, H., Sammori, T., & Kawanami, A. (2000b). The effects of rockfall volume on runout distance. *Engineering Geology, 58*(2), 109–124.

Okura, Y., Kitahara, H., Ochiai, H., Sammori, T., & Kawanami, A. (2002). Landslide fluidization process by flume experiments. *Engineering Geology, 66*(1), 65–78.

Pal, S., Kaynia, A.M., Bhasin, R.K., and Paul, D.K. (2011). Earthquake Stability Analysis of Rock Slopes: a Case Study. *Rock Mechanics and Rock Engineering*.

Papaloizou, L., & Komodromos, P. (2009). Planar investigation of the seismic response of ancient columns and colonnades with epistyles using a custom-made software. *Soil Dynamics and Earthquake Engineering, 29*(11–12), 1437–1454.

Papantonopoulos, C., Psycharis, I. N., Papastamatiou, D. Y., Lemos, J. V., & Mouzakis, H. P. (2002). Numerical prediction of the earthquake response of classical columns using the distinct element method. *Earthquake Engineering & Structural Dynamics, 31*(9), 1699–1717.

Pekau, O. A., & Cui, Y. (2004). Failure analysis of fractured dams during earthquakes by DEM. *Engineering Structures, 26*(10), 1483–1502.

Pfeiffer, T. J., & Bowen, T. (1989). Computer simulation of rockfalls. *Bulletin of the Association of Engineering, 26*(1), 135–146.

Pirulli, M. (2005). Numerical modelling of landslide runout. *A continuum mechanics approach*, Politecnico di Torino.

Poisel, R., Preh, A., & Hungr, O. (2008). Run out of landslides-continuum mechanics versus discontinuum mechanics models. *Geomechanics and Tunnelling, 1*(5), 358–366.

Prevost, J.H. (1981). DYNA-FLOW: a nonlinear transient finite element analysis program. Princeton University, Department of Civil Engineering, School of Engineering and Applied Science.

Psycharis, I., Lemos, J., Papastamatiou, D., Zambas, C., & Papantonopoulos, C. (2003). Numerical study of the seismic behaviour of a part of the parthenon pronaos. *Earthquake Engineering & Structural Dynamics, 32*(13), 2063–2084.

Pyke, R. (1991). Selection of Seismic Coefficients for Use in Pseudo-Static Slope Stability Analyses. http://www.tagasoft.com/Discussion/article2_html

Rathje, E. M., & Bray, J. D. (1999). An examination of simplified earthquake-induced displacement procedures for earth structures. *Canadian Geotechnical Journal, 36*(1), 72–87.

Rathje, E. M., & Bray, J. D. (2000). Nonlinear coupled seismic sliding analysis of earth structures. *Journal of Geotechnical and Geoenvironmental Engineering, 126*(11), 1002–1014.

Richards, R., & Elms, D. G. (1979). Seismic behavior of gravity retaining walls. *Journal of the Geotechnical Engineering Division, 105*(4), 449–464.

Richards, J., Elms, D., & Budhu, M. (1993). Seismic bearing capacity and settlements of foundations. *Journal of Geotechnical Engineering, 119*(4), 662–674.

Rickenmann, D. (1999). Empirical relationships for debris flows. *Natural Hazards, 19*(1), 47–77.

Rodriguez, C.E., Bommer, J., & Chandler, R.J. (1999). Earthquake-induced landslides 1980–1997. *Soil Dynamics and Earthquake Engineering, 18*, 325–346.

Sarma, S. K. (1975). Seismic stability of earth dams and embankments. *Geotechnique, 25*(4), 743–761.

Sarma, S. K. (1981). Seismic displacement analysis of earth dams. *Journal of the Geotechnical Engineering Division, 107*(12), 1735–1739.

Sasaki, T., Hagiwara, I., Sasaki, K., Yoshinaka, R., Ohnishi, Y., & Nishiyama, S. (2004). Earthquake response analysis of rock-fall models by discontinuous deformation analysis. In *In Proceedings of the ISRM international symposium 3rd ARMS* (pp. 1267–1272). Kyoto, Japan.

Sasaki, T., Hagiwara, I., Sasaki, K., Ohnishi, Y., & Ito, H. (2007). Fundamental studies for dynamic response of simple block structures by DDA. In *In Proceedings of the eighth international conference on analysis of discontinuous deformation: fundamentals and applications to mining & civil engineering* (pp. 141–146). Beijing, China.

Sassa, K. (1988). Motion of Landslides and Debris Flows: Prediction of Hazard Area: Report for Grant-in-aid for Scientific Research by Japanese Ministry on Education, Science and Culture (project No. 61480062). Disaster Prevention Research Institute.

Savage, S., & Hutter, K. (1989). The motion of a finite mass of granular material down a rough incline. *Journal of Fluid Mechanics, 199*(1), 177–215.

Sawada, T., Chen, W. F., & Nomachi, S. G. (1993). Assessment of seismic displacements of slopes. *Soil Dynamics and Earthquake Engineering, 12*, 357–362.

Saygili, G., & Rathje, E. M. (2009). Probabilistically based seismic landslide hazard maps: an application in southern California. *Engineering Geology, 109*(3), 183–194.

Scheidegger, A. E. (1973). On the prediction of the reach and velocity of catastrophic landslides. *Rock Mechanics, 5*(4), 231–236.

Seed, H.B. 1973. Analysis of the Slides in the San Fernando Dams During the Earthquake of Feb. 9. (1971). *Report to State of California Department of Water Resources, Los Angeles*

Department of Water and Power. National Science Foundation: College of Engineering, University of California.

Seed, H. B. (1979). Considerations in the earthquake-resistant design of earth and rockfill dams. *Geotechnique, 29*(3), 13–41.

Seed, H.B., & Martin, G.R. (1966). The seismic coefficient in earth dam design. *Journal of Soil Mechanics & Foundations Div, 92*(Proc. Paper 4824).

Serff, N. (1976). *Earthquake induced deformations of earth dams*. College of Engineering, University of California.

Shi, G.-H. (1988). *Discontinuous deformation analysis a new numerical model for the statics and dynamics of block systems*. Berkeley: University of California.

Shi, G. (2002). Single and multiple block limit equilibrium of key block method and discontinuous deformation analysis. In *Proceedings of the 5th International Conference on Analysis of Discontinuous Deformation* (pp. 3–43). Rotterdam: AA Balkema.

Shi, G.-H., & Goodman, R. E. (1985). Two dimensional discontinuous deformation analysis. *International Journal for Numerical and Analytical Methods in Geomechanics, 9,* 541–556.

Shi, G.-H., & Goodman, R. E. (1989). Generalization of two-dimensional discontinuous deformation analysis for forward modelling. *International Journal for Numerical and Analytical Methods in Geomechanics, 13,* 359–380.

Sousa, J., & Voight, B. (1991). Continuum simulation of flow failures. *Geotechnique, 41*(4), 515–538.

Stamatopoulos, C. (1996). Sliding system predicting large permanent co-seismic movements of slopes. *Earthquake Engineering & Structural Dynamics, 25*(10), 1075–1093.

Stewart, J. P., Blake, T. F., & Hollingsworth, R. A. (2003). A Screen Analysis Procedure for Seismic Slope Stability. *Earthquake Spectra, 19*(3), 697.

Taiebat, M., Kaynia, A. M., & Dafalias, Y. F. (2011). Application of an anisotropic constitutive model for structured clay to seismic slope stability. *Journal of Geotechnical and Geoenvironmental Engineering, 137*(5), 492.

Takahashi, T., Momiyama, A., Hirai, K., Hishinuma, F., & Akagi, H. (1992). Functional correlation of fetal and adult forms of glycine receptors with developmental changes in inhibitory synaptic receptor channels. *Neuron, 9*(6), 1155–1161.

Terzaghi, K. (1950). *Theoretical Soil Mechanics*.

Tsesarsky, M., Hatzor, Y., & Sitar, N. (2005). Dynamic displacement of a block on an inclined plane: analytical, experimental and DDA results. *Rock Mechanics and Rock Engineering, 38* (2), 153–167.

Varnes, D.J. (1984). Landslides, t.I.A.E.G.C.o., and Slopes, O.M.M.o. *Landslide hazard zonation: a review of principles and practice*.

Wang, F., Sun, P., Highland, L., & Cheng, Q. (2012). Initiation and motion mechanism of the Donghekou rapid and long runout landslide triggered by the 2008 Wenchuan earthquake, China. In *The International Symposium on Earthquake-induced landslides* (pp. 473–483). Kiryu, Japan: Springer-Verlag Berlin Heidelberg.

Wartman, J., Asce, M., Bray, J. D., & Seed, R. B. (2003). Inclined plane studies of the newmark sliding block procedure. *Journal of Geotechnical and Geoenvironmental Engineering, 129*(8), 673–684.

Wasowski, J., Keefer, D. K., & Lee, C.-T. (2011). Toward the next generation of research on earthquake-induced landslides: current issues and future challenges. *Engineering Geology, 122* (1–2), 1–8.

Wilson, R. C., & Keefer, D. K. (1983). Dynamic analysis of a slope failure from the 6 August 1979 Coyote Lake, California, earthquake. *Bulletin of the Seismological Society of America, 73* (3), 863–877.

Wu, J.-H. (2003). *Numerical analysis of discontinuous rock masses using discontinuous deformation analysis*. Kyoto, Japan: Kyoto University.

Wu, J.-H. (2010). Seismic landslide simulations in discontinuous deformation analysis. *Computers and Geotechnics, 37*(5), 594–601.

Wu, J.-H., & Chen, C.-H. (2011). Application of DDA to simulate characteristics of the Tsaoling landslide. *Computers and Geotechnics, 38*(5), 741–750.

Wu, J.-H., & Tsai, P.-H. (2011). New dynamic procedure for back-calculating the shear strength parameters of large landslides. *Engineering Geology.*

Wu, A., Ren, F., & Dong, X. (1997). A study on the numerical model of DDA and its preliminary application to rock engineering. *Chinese Journal of Rock Mechanics and Engineering, 16*(5), 411–417.

Wu, J., Lin, J., & Chen, C. (2009). Dynamic discrete analysis of an earthquake-induced large-scale landslide. *International Journal of Rock Mechanics and Mining Sciences, 46*(2), 397–407.

Yagoda-Biran, G., & Hatzor, Y. H. (2010). Constraining paleo PGA values by numerical analysis of overturned columns. *Earthquake Engineering & Structural Dynamics, 39*(4), 463–472.

Yegian, M.K. (1991). *Seismic risk analysis for earth dams.* ASCE.

Yegian, M. K., Marciano, E. A., & Ghahraman, V. G. (1991). Earthquake-induced permanent deformations: probabilistic approach. *Journal of Geotechnical Engineering, 117*(1), 35–50.

Yegian, M., Harb, J., & Kadakal, U. (1998). Dynamic response analysis procedure for landfills with geosynthetic liners. *Journal of Geotechnical and Geoenvironmental Engineering, 124* (10), 1027–1033.

Zhang, C., Pekau, O.A., Jin, F., & Wang, G. (1997). Application of distinct element method in dynamic analysis of high rock slopes and blocky structures. *Soil Dynamics and Earthquake Engineering, 16*, 385–394.

Zhang, Y., Chen, G., Zheng, L., Wu, J., & Zhuang, X. (2012a). Effects of vertical seismic force on the initiation of the Da-guangbao landslide induced by the Wenchuan earthquake. In *The 8th Annual Conference of International Institute for Infrastructure, Renewal and Reconstruction* (pp. 530–539). Kumamoto, Japan.

Zhang, Y., Chen, G., Zheng, L., & Li, Y. (2012b). Numerical analysis of the largest landslide induced by the Wenchuan earthquake, May 12, 2008 using DDA. In *International Symposium on Earthquake-induced Landslides*. Kiryu, Japan.

Zhang, Y., Chen, G., Zheng, L., Li, Y., & Zhuang, X. (2013). Effects of geometries on three-dimensional slope stability. *Canadian Geotechnical Journal, 50*(3), 233–249.

Zhao, S.L., Salami, M.R., & Rahman, M.S. (1997). Discontinuous De-formation Analysis Simulation of Rock Slope Failure Processes. In *9th International Conference on Computer Methods and Advances in Geomechanics* (pp. 473–477), Wuhan, China.

Zheng, L. (2010). *Development of new models for landslide simulation based on discontinuous deformation analysis.* Fukuoka, Japan: Kyushu University.

Zheng, Y. R., Tang, X., Zhao, S., Deng, C., & Lei, W. (2009). Strength reduction and step-loading finite element approaches in geotechnical engineering. *Journal of Rock Mechanics and Geotechnical Engineering, 1*(1), 21–30.

Chapter 3
Seismic Slope Stability Analysis by Considering Tension Crack

Abstract This chapter proposes an approach for how to consider tension failure in addition to shear failure in widely used FLAC^{3D}, a finite difference method, so that stability analysis can also be carried out for a slope with a complex slip surface. It is shown that the safety factors estimated using the numerical methods are almost the same comparing to an analytical method for a homogeneous slope. Additionally, it is shown by a large number of analysis examples that the effect of tension failure on slope stability is significant and the safety factor will be incorrect if tensile failure is ignored in the case of seismic loading.

Keywords Slope stability · FLAC^{3D} · Tension failure

3.1 Introduction

Terzaghi, who is considered by many as the father of geotechnical engineering, has said:

> …the questionable action and the questionable depth of the tension zone have considerable bearing on the limited dependability of many stability analyses…

at his classical works, theoretical soil mechanics (Terzaghi 1950. Indirectly refer to Utili 2013). Unfortunately, very little progress has been made since Terzaghi's time in the assessment of the influence of tension crack on the slope stability, although cracks are often appeared in the top surface of cohesive soils and rock slopes. Previous literatures investigating the influence of tension crack on slope stability analysis involve mainly the use of limit equilibrium methods in their classical form, e.g., Spencer (1968), Kaniraj and Abdullah (1993), or are based on variational formulations, e.g., Baker (1981, 2003). Other works investigated tension cracks in undrained conditions, e.g., Baker and Leshchinsky (2001). More recently, finite element upper-bound limit analyses have been attempted, where the presence of a crack of specified location and depth is included in the geotechnical analysis of a sheet pile (Antao et al. 2008).

Y. Zhang, *Earthquake-Induced Landslides*, Springer Natural Hazards,
DOI 10.1007/978-981-10-2935-6_3

Most slope stability analyses do not include the process of tension crack propagation. At most, a preexisting crack is considered through modification of geometry in analyses. There are two reasons for this situation: (i) It is very difficult for analytical methods or numerical simulation methods to calculate the collapse value of the slope with an undetermined discontinuity introduced by the tension crack and (ii) comparing to the difficulty in the calculation, influence of tension crack on slope is relatively little and ignored or preset by most analyses.

However, when a slope affected by an earthquake, the influence of tension failure may too significant to be ignored or preset. The frequently occurring earthquake events in recent years in New Zealand, Japan, and China have led to a renewed knowledge in mechanism of instability of slopes. A large number of evidences from investigations of earthquake-induced landslides and shaking table test show that significant tension failures appear in top of almost all landslides or potential sliding slopes. Figure 3.1 gives some cases of earthquake-induced landslides with significant tension failure. Figure 3.1a, b shows steep scarps with coarse crack of Donghekou landslide and Pingxicun landslide induced by 2008 Wenchuan

Fig. 3.1 Tension failures in earthquake-induced landslides. Steep scarp with coarse cracks of **a** Donghekou landslide, Qingchuan and **b** Pingxicun landslide, Pingwu; **c** Tension cracks in the top of Shiziliang (Photos from Xu) and **d** Tension crack observed in shaking table test (from Wakai et al. 2001)

earthquake, and Fig. 3.1c shows tension crack appeared in a potential landslide and a result of shaking table test carried out by Wakai et al. (2001) in which tension crack can be obviously found in top of slope shown in Fig. 3.1d.

There are a large number of described, analytical, and numerical studies which have provided supporting evidences of the significant influence of tension failure on slope stability analysis. Huang (2009), Huang et al. (2011), Xu et al. (2009), and Yin et al. (2009) have gave much detailed description for tension segment of slope failure surface based on post-earthquake investigations. Zheng et al. (2009) and Zhang et al. (2012a, b) have certificated the certain influence of tension failure zone by analyzing the mechanism of seismic slope using numerical simulation. Zhang et al. (2011a, b) showed significant effect of tension failure on slope stability analysis and initiation velocity using analytical methods. These studies, however, mostly focus on the description or explanation of phenomenon, few on deep research of stability analysis subjected to the tension failure.

From simple static considerations, it is apparent that a crack in a uniform cohesive-frictional (c-φ) slope must depart from the ground surface along a vertical line. It can be expected that cracks may deviate from the vertical as they go deeper, because of the rotation of the minor tensile principal stress. However, this study concerns only with vertical cracks, since this constitutes a first step into the understanding of the problem, and very little is known about non-vertical crack patterns in cohesive soils, with all the relevant literatures on the subject confined to the case of vertical cracks. According to some lower-bound analyses (e.g., Terzaghi 1950; Baker 2003; Antao et al. 2008), which assume a limited tensile strength for the soil, the maximum crack depth is limited. However, recent post-earthquake investigations show that the cracks are often larger than the theoretical value. Those cracks are caused and/or deepened by cycles of seismic loading. In light of these considerations, cracks of any possible depth and location have been investigated in this study. The reader interested in tension cracks only will have to calculate a limit depth of interest according to a known or assumed tensile resistance for the geo-material and to an analysis of the local state of stress. The depth of the cracks could be calculated from a static analysis, by means of either analytical or numerical methods (e.g., finite element method). It is worth noting that the results obtained in this study are applicable not only to geomaterials of no-tensile strength, but also to any cohesive-frictional geomaterial (soil or rock) whose tensile strength is known. However, in most cases, accurate estimates of crack depths are not available: Therefore, the stability of a slope needs to be analyzed for a range of depths rather than for a single value.

Therefore, to investigate how the tension failure effects on seismic slope stability analysis, at first, numerical simulation method of FDM is proposed by considering a tension crack in Sect. 3.2, and then, comparison and discussion are given in Sects. 3.3 and 3.4, respectively.

3.2 Numerical Simulation Method

As stated previously, pseudo-static method can be simply and directly used to identify the FOS of a slope, but it cannot simulate the transient dynamic effects of earthquake shaking, because it assumes a constant unidirectional pseudo-static acceleration. Therefore, in particular, it is recommended that pseudo-static methods are used only for preliminary assessments and screening procedures, which then should be followed by more sophisticated analysis. Stress–strain method has been used increasingly in engineering practice and becomes more and more popular for the real dynamic analysis with the development of computer technology and simulation approach. In these numerical simulation methods, FLAC^{3D}, a finite difference method, is a widely used method. Therefore, FLAC^{3D} is used herein to simulate the seismic slope.

3.2.1 *Failure Mechanism in the FLAC^{3D}*

At present, the failure mechanism of seismic slope follows the static slope failure mechanism, i.e., the main reason that caused seismic slope instable is shear failure while ignoring the influence of tension failure. In fact, with the reason of small tensile strength and the action of earthquake loading, the slope in reciprocating motion is more easily to be tensioned. Stress states of a point in slope mass at two situations, static and dynamic, are illustrated as in Fig. 3.2. As the description of most soil mechanics textbooks, compressive stresses are considered positive, and tensile stresses are negative. The major and minor principal stresses σ_1 and σ_3 of static state are obtained from

$$\sigma_1 = \frac{\sigma_x + \sigma_z}{2} + \sqrt{(\frac{\sigma_x - \sigma_z}{2})^2 + \tau_{xz}^2} \tag{3.1}$$

$$\sigma_3 = \frac{\sigma_x + \sigma_z}{2} - \sqrt{(\frac{\sigma_x - \sigma_z}{2})^2 + \tau_{xz}^2} \tag{3.2}$$

where σ_x, σ_z are stresses in horizontal direction and vertical direction, respectively, and τ_{xz} is shear stress.

A shear stress τ_s caused by the earthquake loading can be added into the existing stress state when the slope is effected by a horizontal earthquake loading which propagates from bottom upward to top of the slope. The magnitude and direction of dynamic stress τ_s are time-dependent. If not consider the influence of wave reflection and refraction, the stress state in this situation can be simply indicated as in Fig. 3.2. The combined major and minor principal stresses σ_1', σ_3' are calculated from

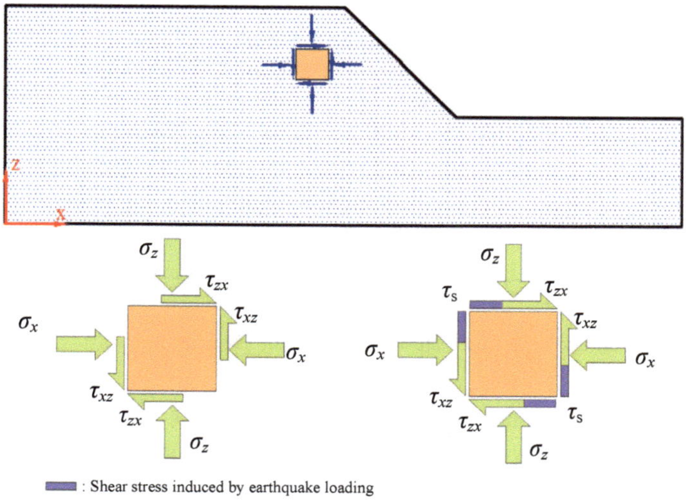

Fig. 3.2 Stress state of a point in slope mass at conditions of **a** static and **b** horizontal seismic shear stress loadings

$$\sigma_1' = \frac{\sigma_x + \sigma_z}{2} + \sqrt{\left(\frac{\sigma_x - \sigma_z}{2}\right)^2 + (\tau_{xz} + \tau_s)^2} \tag{3.3}$$

$$\sigma_3' = \frac{\sigma_x + \sigma_z}{2} - \sqrt{\left(\frac{\sigma_x - \sigma_z}{2}\right)^2 + (\tau_{xz} + \tau_s)^2} \tag{3.4}$$

where τ_s obtained from:

$$\tau_s = \rho C_s v_s \tag{3.5}$$

where ρ is mass density of soil material, C_s is speed of s-wave propagation through material, and v_s is input shear particle velocity caused by earthquake loading in horizontal direction. Equations 3.3 and 3.4 show an opposite change pattern of major and minor principal stresses σ_1', σ_3'. In the (σ, τ) plane, this change pattern is presented as expand or narrow of the Mohr circle (Fig. 3.3).

Note that if the influence of wave reflection and refraction is considered, the expressions of major and minor principal stresses σ_1' σ_3' will be very complex. This point will be described in detail by monitoring records through the time history of seismic excitations.

Many failure criterions have been presented for modeling the strength of soil, and the Mohr–Coulomb criterion remains the one most widely used in geotechnical practice. A modified Mohr–Coulomb failure criterion is used in this study. The

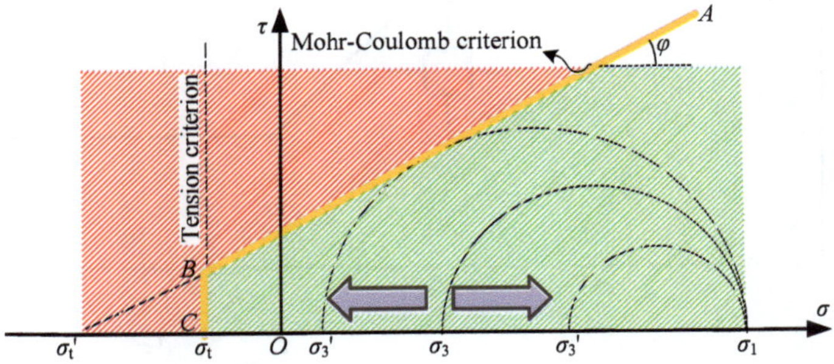

Fig. 3.3 Failure criterion combined tension-shear failure mechanism

representation of the failure criterion in the (σ, τ) plane is sketched in Fig. 3.3. The failure envelope is defined from point A to B by the Mohr–Coulomb yield function

$$f_{\text{shear}} = \frac{\sigma_1' - \sigma_3'}{2} - c\cos\varphi - \frac{\sigma_1' + \sigma_3'}{2}\sin\varphi = 0 \quad (3.6)$$

and from B to C by the tension yield function

$$f_{\text{tension}} = \sigma_3' - \sigma_t = 0 \quad (3.7)$$

where c and φ are cohesion and friction angle of soil material, and σ_t is tension strength of soil material.

Because tensile strength of soil material is much smaller than their shear strength, we can take the opinion that once the minor principal stress σ_3 equals to tensile strength σ_t, the tension failure will occur and the tension failure surfaces parallel to minor principal stress plane.

The failure functions f_{tension} and f_{shear} can be interpreted as follows

- $f_{\text{tension}} > 0$ and $f_{\text{shear}} < 0$ note that the normal and shear stresses on a plane in a soil mass inside failure envelope, green zone in Fig. 3.3. Failure will not occur in this situation;
- $f_{\text{tension}} = 0$ or $f_{\text{shear}} = 0$ notes that the normal and shear stresses on a plane in a soil mass on failure envelope, yellow line in Fig. 3.3. The material is yielding state in this situation.
- $f_{\text{tension}} < 0$ or $f_{\text{shear}} > 0$ notes a state of stress plotted as red zone in Fig. 3.3 that cannot exist; since the stresses outside failure envelope, failure would have occurred before this condition was reached.

For a slope stability analysis problem, tensile failure should be first considered, because tension strength of soil material is usually much smaller than shear strength. If taking *cut-through* of the tension failure zone and shear failure surface as definition of slope failure, compute process can be illustrated as in Fig. 3.4.

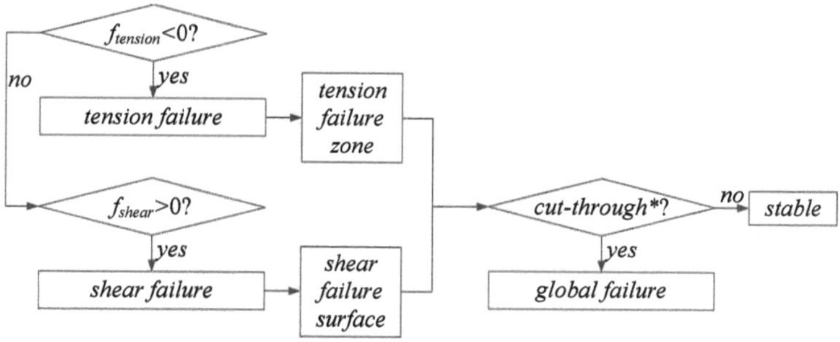

*: Assume take *cut-through* of tension failure zone and shear failure surface as definition of slope failure

Fig. 3.4 Flowchart of seismic slope stability analysis based on tension-shear failure mechanism

Definition of slope failure is discussed in the third subsection. Note that the procedure of tension failure and shear failure is automatically included in FLAC^{3D} program, while the global failure needs own judgment based on different definitions.

3.2.2 Dynamic Formulation

3.2.2.1 Brief Description of Dynamic Analysis Using FLAC^{3D}

The program used in this paper is finite difference program FLAC^{3D} version 3.10 (Itasca 2007). The calculation is based on the explicit finite difference scheme to solve the full equations of motion, using lumped grid point masses derived from the real density of surrounding zones rather than fictitious masses used for optimum convergence in the static solution scheme. And a flowchart of dynamic analysis for a slope is illustrated in Fig. 3.5.

3.2.2.2 Modeling with FLAC^{3D}

In the whole process, there are several important aspects should be considered while preparing a FLAC^{3D} model for dynamic analysis: in terms of (1) dynamic loading, (2) boundary conditions, (3) mechanical damping, and (4) wave transmission through the model. They are discussed in this section.

This section analyzes the critical example studied in last section. The homogeneity slope has the height of 20 m and the incline angle of 45° that has studied by some earlier researchers (e.g., Zheng et al. 2009; Zhang et al. 2012a). A schematic illustration of the 2D analyzed mesh and the boundary conditions is provided in Fig. 3.6. In order to study the development process of slope failure surface, one square meter of grid size is meshed. Both x and y displacements are fixed at the base

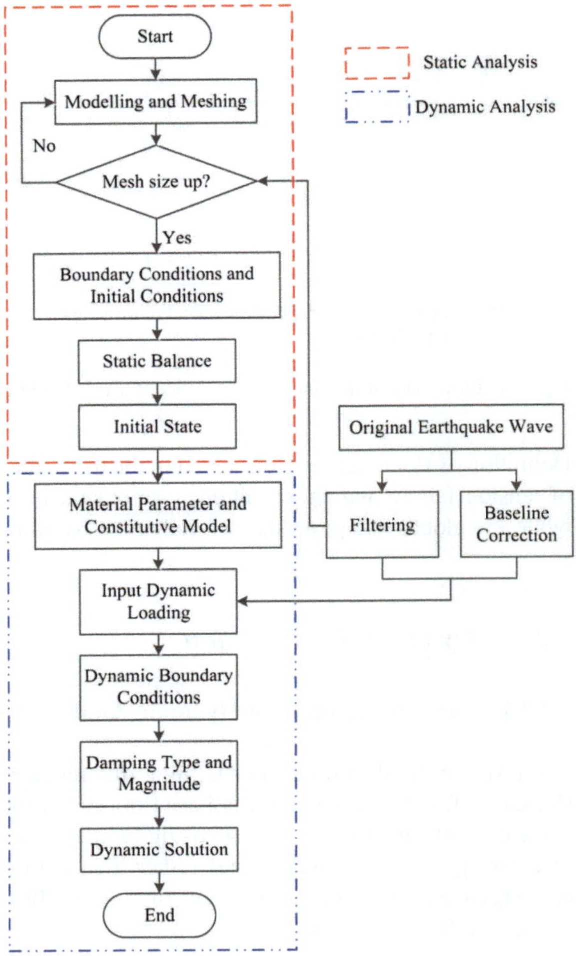

Fig. 3.5 Flowchart of seismic slope stability analysis using FLAC^{3D} program (Modified from Chen and Xu 2008)

of the model. And x displacements are fixed on either side of the model along the y-axis. The slope is allowed free to move in both the directions. Free field boundary is used in the present model to minimize the wave reflection. This boundary condition provides a better representation when quiet boundaries are used in conjunction with external seismic sources. The size of slope ensures the assumption of free field boundary. For a dynamic analysis, FLAC^{3D} version 3.10 program provides several mechanical damping in which local damping is a simple and pragmatic method. The local damping coefficient α_L is defined as

$$\alpha_L = \pi D \tag{3.8}$$

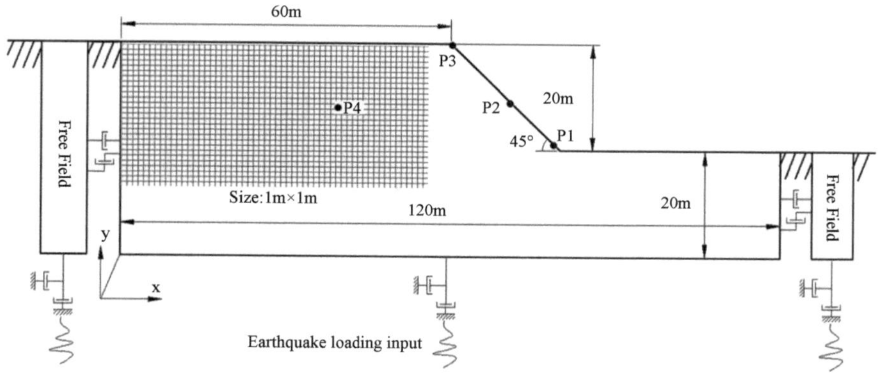

Fig. 3.6 Mesh generation and boundary conditions of finite difference model for dynamic slope stability analysis

where D is fraction of critical damping. Although the actual value given to the local damping has a profound influence on the dynamic wave transmission, if it is chosen from a certain range, it has little influence on the predicted factor of safety in seismic slope stability analysis. Hence, local damping of 0.157 (i.e., fraction of critical damping is 5%) is used in the model as suggested by other studies for these kinds of problems.

Table 3.1 shows the property parameters of soil material used in the analysis. It should be noted that some unlisted parameters, e.g., unit weight γ, modulus of volume K, and shear modulus G, are also used in this paper. They can be translated from the parameters listed in Table 3.1. The density ρ assigned to the soil decides the total unit weight γ and to move forward the self-gravity load. The parameters c and φ note the effective cohesion and interfriction angle of the soil material. The dilation angle ψ affects the volume change of the soil during yielding. In this study, we take $\psi = \varphi$, i.e., the plasticity flow rule is associated and direct comparisons with theorems from classical plasticity can be made. In addition, it should be noted that the tensile strength σ_t is taken as 0, because it is so unreliable that can be ignored.

Table 3.1 Property parameters of slope material

Parameters	Value
Modulus of elasticity E	77.48×10^6 Pa
Poisson's ratio μ	0.3
Density ρ	2000 kg/m^3
Cohesion c	40,000 Pa
Interfriction angle φ	20°
Dilation angle ψ	20°
Tensile strength σ_t	0

3.2.2.3 Earthquake Loading

(i) **Original earthquake loading**

The dynamic load applied here is the transverse component of the acceleration time history (record name: KJM-0°) modified from the Kobe earthquake, occurred in Japan, 1997, as shown in Fig. 3.10. The total duration of the earthquake loading is 15 s with a time step of 0.02 s. The $a_{\max}$ value of the recorded earthquake is 0.2046 g at time of 3.52 s. The earthquake loading is applied at bottom of the whole stratum. From the acceleration time history record, velocity and displacement time histories can be computed by once and twice integration, respectively.

(ii) **Baseline correction**

If the acceleration record shown in Fig. 3.7 is directly used as an earthquake loading time history, the FLAC^{3D} model may exhibit continuing velocity or residual displacements after the motion has finished as shown in Fig. 3.8a, b. This arises from the fact that the integral of the complete acceleration time history may not be zero. In this paper, a low-frequency velocity wave which, when added to the original history, produces a final displacement which is zero is used. The action will not affect the mechanics of the deformation of the model. The process of baseline correction and corrected displacement time histories are shown in Fig. 3.8.

(iii) **Translation of stress time history**

In order to avoid the effect of the quiet boundary be nullified, a stress boundary condition is used. This stress history is transformed from a velocity record using the formula

$$\sigma_s = 2(\rho C_s)v_s \tag{3.9}$$

where C_s is the speed of s-wave propagation through medium; v_s is input shear particle velocity time history of earthquake loading. The factor of two in the

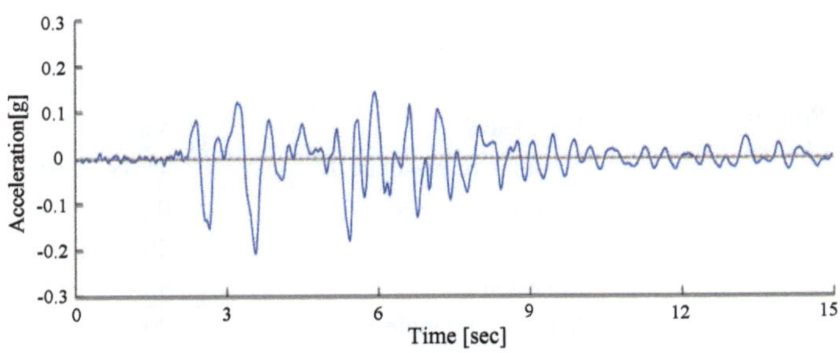

Fig. 3.7 Acceleration time history of earthquake loading applied in study

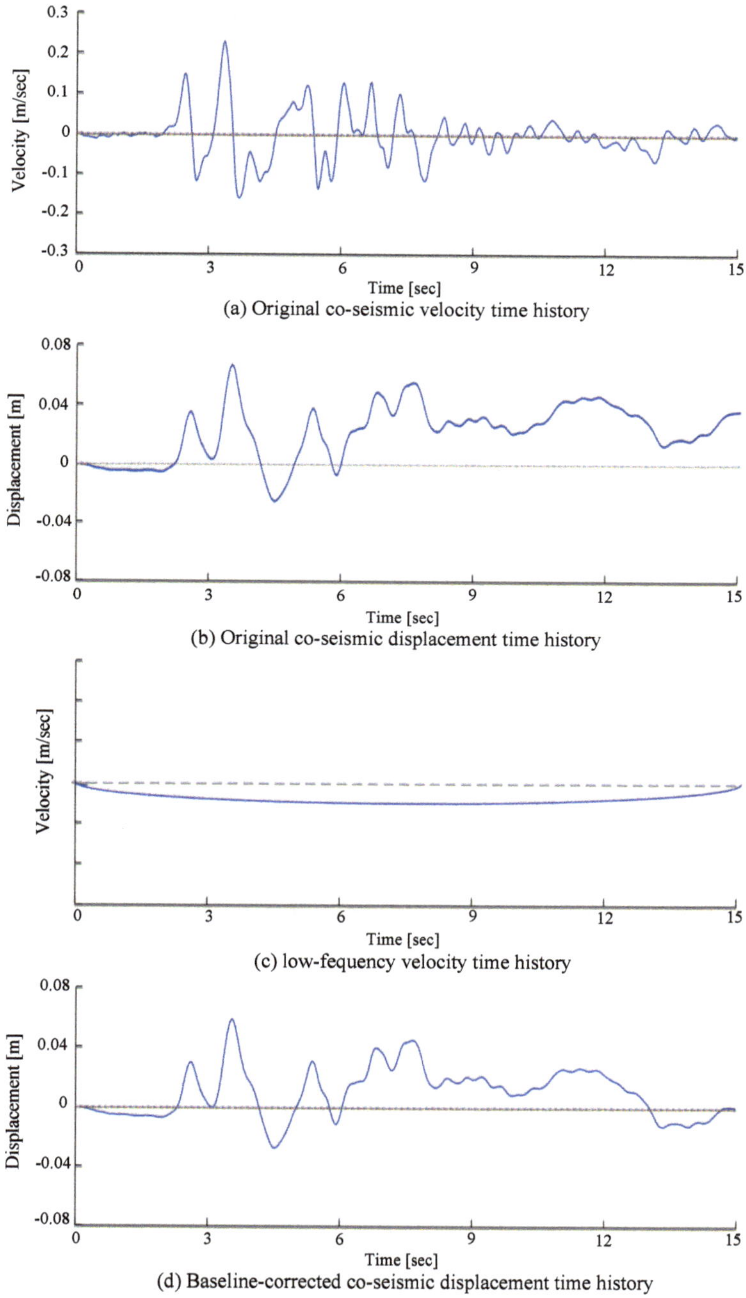

Fig. 3.8 Baseline correction process

equation accounts for the fact that the applied stress must be doubled to overcome the effect of the viscous boundary. C_s is given by

$$C_s = \sqrt{G/\rho} \tag{3.10}$$

where G is the shear modulus of material and given by

$$G = E/[2(1+\mu)] \tag{3.11}$$

Figure 3.9 shows the input stress time history shifted from velocity time history.

3.2.3 Results

In this section, three major results of seismic slope stability analysis are presented. The first is the validation of stress state and relationship with tension failure. The

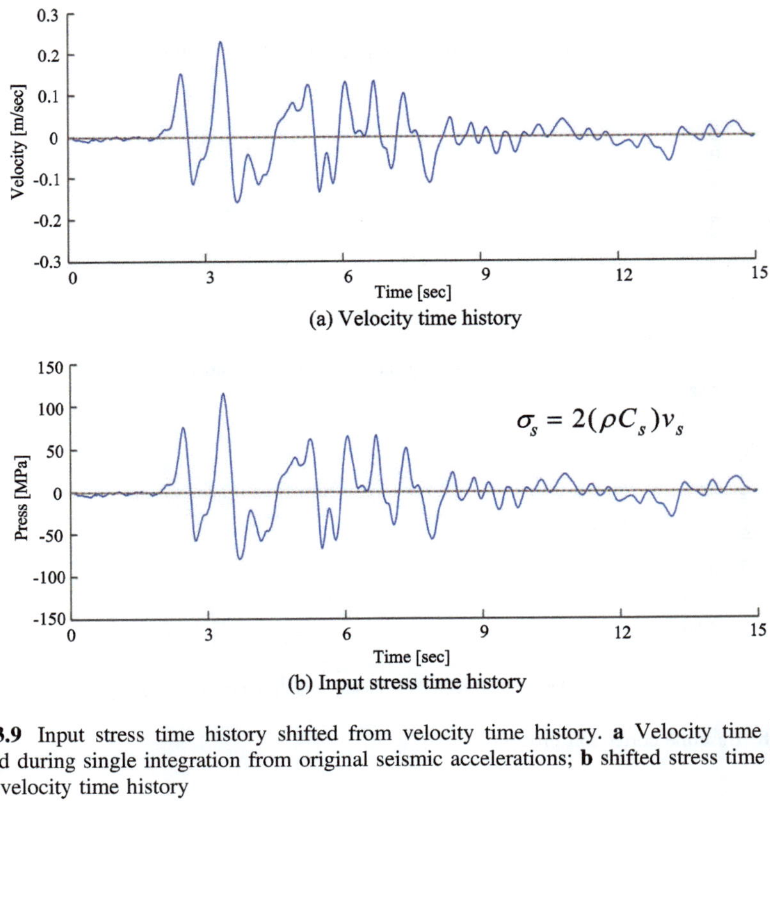

Fig. 3.9 Input stress time history shifted from velocity time history. **a** Velocity time history shifted during single integration from original seismic accelerations; **b** shifted stress time history from velocity time history

second is about the permanent displacement in actual strength case. The third is the factor of safety under different failure mechanisms, and the forth aspect is the shape of slip surface.

3.2.3.1 Stress and Tension Failure

Figure 3.10 shows the major and minor prince stress records of a monitoring point set away from the slope surface and with a certain depth from top surface in where the influence of wave reflection and refraction is small. From the figure, an opposite change pattern of major and minor principal stresses σ_1' σ_3' can be found that consisted with the description of Sect. 3.2.1.

As previous yield criterion, tension-shear failure criterion, tension failure is firstly judged based on the minor prince stress. The depth of tension failure zone can be judged. Figure 3.11 shows a serious minor prince stress records of points that have different depth from top of slope. As the tension failure criterion, once the minor prince stress smaller than tensile strength of soil, tension failure will occur. In Fig. 3.11, black dash line notes tensile strength of soil material, and stress records

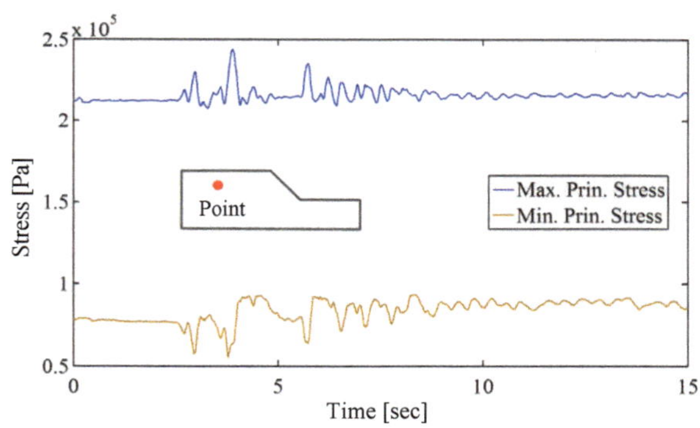

Fig. 3.10 Prince stress records of a point in slope under an earthquake loading

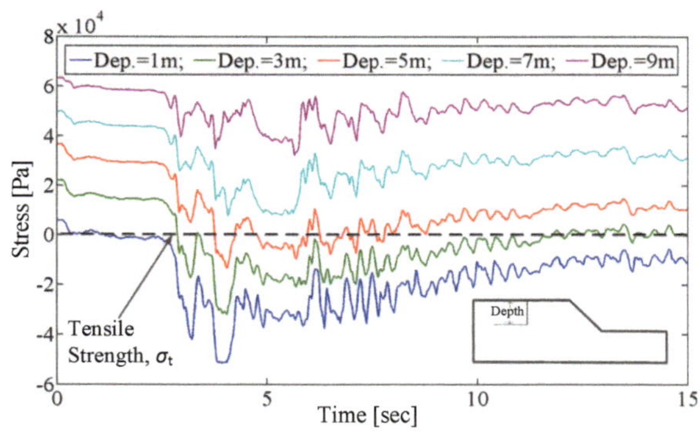

Fig. 3.11 Minor prince stress records of monitoring points

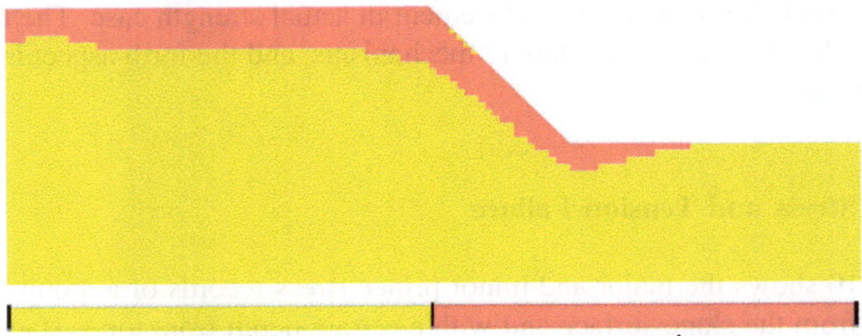

Fig. 3.12 Tension state of slope after the shaking

of points in depth of 1–5 m outstrip the tensile strength line. So we can take the conclusion that depth of tension failure zone is 5–7 m. Figure 3.12 shows the block tension state in which 5–7 m depth of tension failure zone can be found.

3.2.3.2 FOS Based on Tension-Shear Failure Mechanism

As previous mentioned, FOS can be computed by reducing the soil strength in stages to bring the slope into a state of limiting equilibrium. The stability analysis of slope is carried out based on tension-shear failure mechanism.

Take the *cut-through* of tension failure zone and shear failure surface as definition of seismic slope failure, the failure surface contains tension and shear segment. Tension segment is depended on the depth of tension failure zone, and shear segment is decided by the shear strain increment, i.e., cumulant of shear strain. After a series of trying calculation, it can be obtained that *cut-through* of tension plastic zone and shear failure zone occurred in the smallest SRF of 0.99. Figure 3.13 shows the contours of tension plastic zone and shear strain increment at situations of SRF = 0.98 and SRF = 0.99. The maximum values of shear strain increment (SSI_{max}) are 0.03219 and 0.03593, respectively. It can be found that the significant shear strain increment not increases to the tension zone at SRF = 0.98, while *cut-through* of tension failure and shear failure is obtained at SRF = 0.99. Hence, we can take the opinion that the slope is in the limit state at SRF = 0.98. As the previous definition of FOS, it is 0.98 based on the tension-shear failure mechanism.

3.2.4 Discussions

3.2.4.1 Tensile-Shear Failure Mechanism Versus Shear-Only Failure Mechanism

If only considering the shear failure mechanism, failure surface of slope is just induced by the shear failure. Take the *cut-through* of shear plastic zone as definition

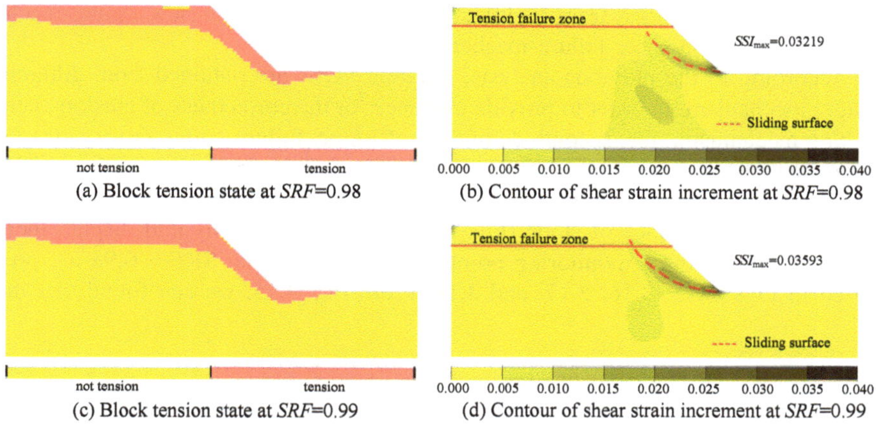

Fig. 3.13 Block tension state and contour of shear strain increment at SRF = 0.98 and SRF = 0.99

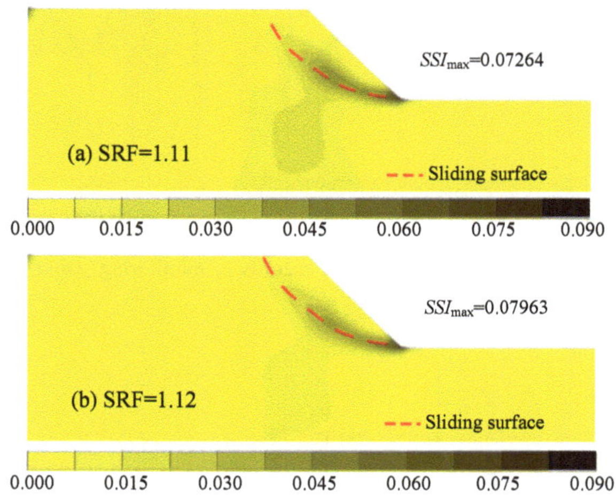

Fig. 3.14 Contours of shear strain increment at **a** SRF = 1.11 and **b** SRF = 1.12

of seismic slope failure as well; after a series of trying calculation, it can be obtained that *cut-through* of shear plastic zone occurred in the smallest SRF of 1.12. Fig. 3.14 shows the contours of shear strain increment at SRF = 1.11 and SRF = 1.12, respectively. The maximum values of shear strain increment (SSI_{max}) are 0.07264 and 0.07963 at SRF = 1.11 and SRF = 1.12, respectively. These results show that the slope is in the limit state at the situation of SRF = 1.11. As the previous definition of factor of safety and the *cut-through* definition of slope failure, we can come to the conclusion that FOS is 1.11 based on single shear failure mechanism.

The FOS of 1.11 calculated based on traditional failure mechanism is larger than that of 0.98 based on tension-shear failure mechanism. For the studied slope here,

FOS reduces about 12% from traditional failure mechanism to new failure mechanism, i.e., tension-shear failure mechanism.

According to the previous analysis, different FOSs are obtained from different failure mechanism. In order to provide evidence for the correctness of tension-shear failure mechanism, the second definition of slope failure, *non-convergence* of permanent displacement of potential failure mass, is used here. Permanent displacements of potential failure mass are recorded by setting some key monitoring points alone the surface of slope (Fig. 3.6). Horizontal velocity and displacement time histories of three monitoring points at SRF = 1.11 and SRF = 0.98 are proposed in Figs. 3.15, 3.16, 3.17, and 3.18. From Fig. 3.15, we can found that the

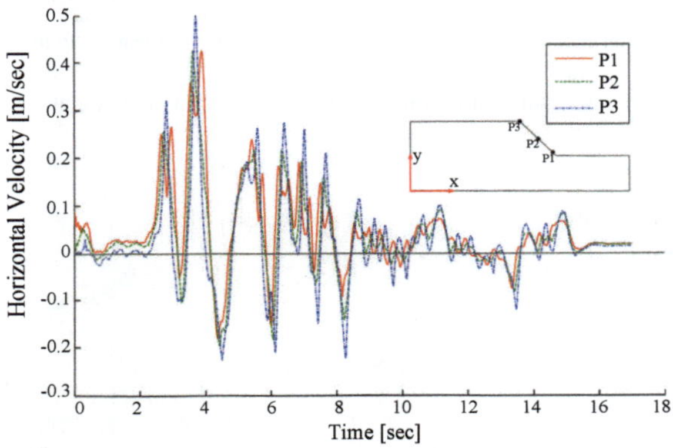

Fig. 3.15 Horizontal velocity records of monitoring points at slope surface at SRF = 1.11

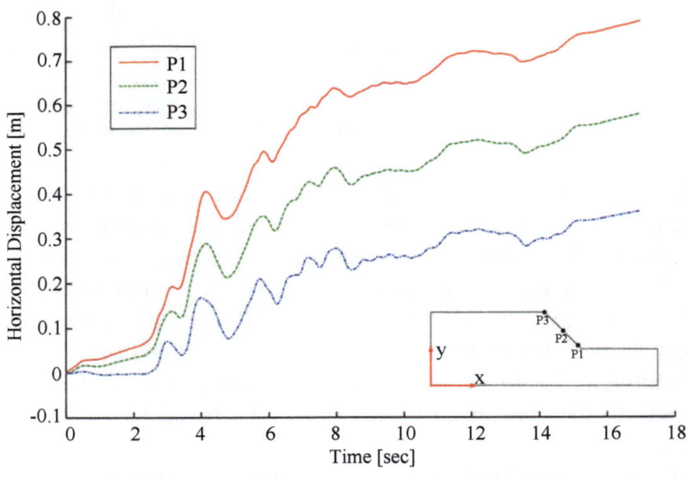

Fig. 3.16 Horizontal displacement records of monitoring points at slope surface at SRF = 1.11

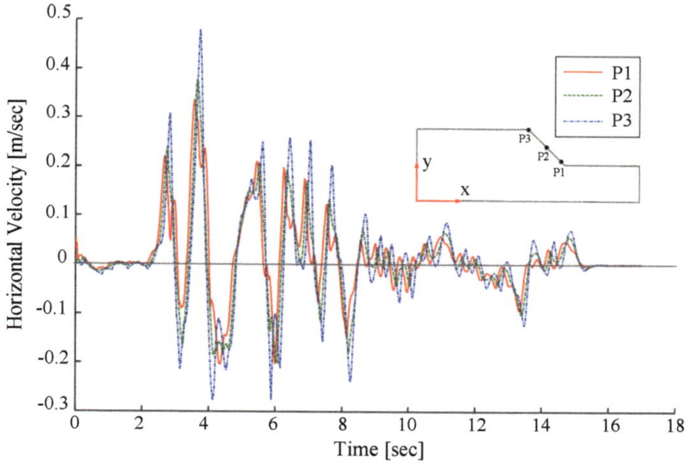

Fig. 3.17 Horizontal velocity records of monitoring points at slope surface at SRF = 0.98

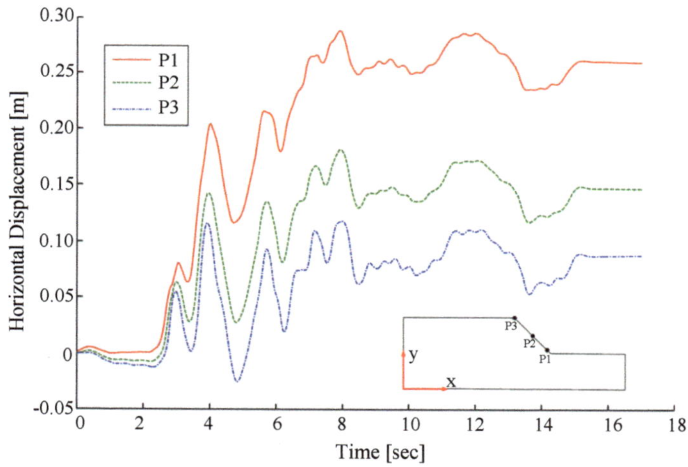

Fig. 3.18 Horizontal displacement records of monitoring points at slope surface at SRF = 0.98

residual velocity of monitoring points after the shaking is not equal to zero at SRF = 1.11, and from Fig. 3.16, *non-convergence* of permanent displacement also be proposed, i.e., the slope is not stable at SRF = 1.11. In opposite, Figs. 3.17 and 3.18 show the velocity that equals to 0 and convergence of permanent displacement, respectively, i.e., the slope is stable at SRF = 0.98. These results confirm the correctness and reasonable of tension-shear failure mechanism.

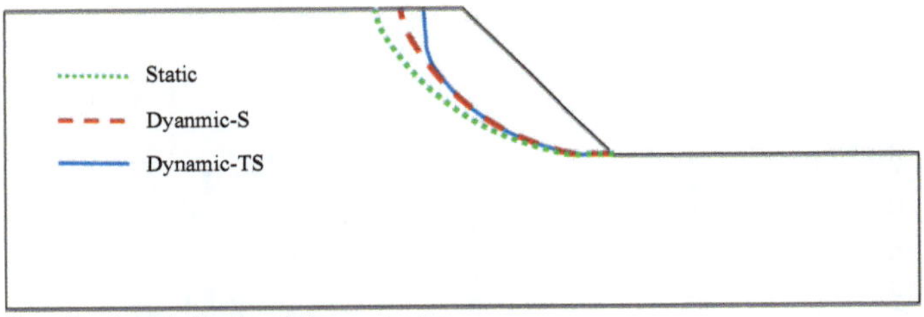

Note: *S* in *dynamic-S* is shear failure mechanism
TS in *dynamic-TS* is tension-shear failure mechanism

Fig. 3.19 Failure surfaces under static and dynamic situations based on different failure mechanisms

3.2.4.2 Shape of Slip Surface

The shape of slip surface is an important aspect of seismic slope stability analysis. It has a close relationship with the volume of failure mass that plays a determination role on the starting velocity and travel distance of a landslide. Figure 3.19 shows the failure surfaces of slope in different cases: static case and dynamic cases based on shear mechanism and tension-shear mechanism. From Fig. 3.19, we can found that in dynamic cases, shapes of slip surface are shallower than those under static case (dense green dotted line). More important, the failure surface under tension-shear failure mechanism (blue solid line) contains two segments: tension failure and shear failure which consist with the result of model test (Wakai et al. 2001; Lin and Wang 2006) and post-earthquake investigation (Huang 2009; Xu et al. 2009).

3.2.4.3 Progressive Failure

The failure progress of slope has an important significance for design of reinforce slope against earthquake loading. Considering the tension-shear failure mechanism, Fig. 3.20 shows the failure progress of seismic slope at different dynamic time. From Fig. 3.20, it can be obviously found that the instability of slope is a progressive failure in which the shear failure zone is expanded slowly from the toe upward to the top of slope and achieves a *cut-through* with tension failure zone. In this example, the slope is not collapsed at dynamic time $t = 3$ s at which the first peak of wave of the earthquake loading goes through. After the largest peak of wave shaking at time of 6 s, the slope has a significant performance, but the cut-through of tension zone and shear failure zone is not achieved. The global instability occurred at time $t = 9$ s at which the large peak of wave has went through. From the figure of tension states of blocks, we can find that the tension failure occurred at the time of largest peak of wave go through. It should be noted that the result presented here is just a tentative results. The research of tension failure mechanism of seismic slope is at the primary stage, and there is still plenty work to do.

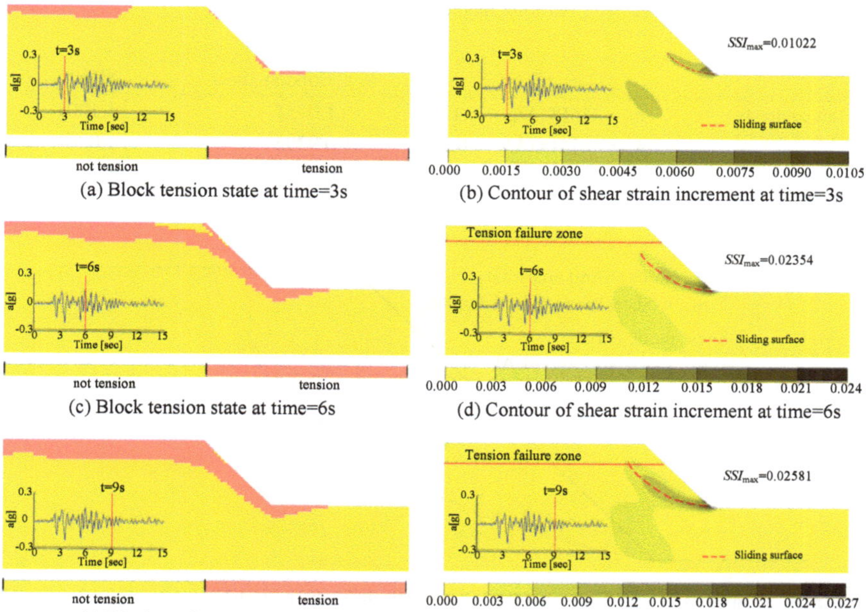

(a) Block tension state at time=3s
(b) Contour of shear strain increment at time=3s
(c) Block tension state at time=6s
(d) Contour of shear strain increment at time=6s
(e) Block tension state at time=9s
(f) Contour of shear strain increment at time=9s

Fig. 3.20 Process of seismic slope failure

3.3 Comparison of Limit Analysis and Numerical Simulation

3.3.1 Comparison of the Depth of Tension Failure

For the studied case, on the one hand, result for the depth of tension failure from limit analysis is 5.8 m depth of tension failure for horizontal seismic coefficient of 0.1–0.15. On the other hand, result from the dynamic simulation from FLAC^{3D} is 5–7 m (Figs. 3.10 and 3.11). Hence, it is shown that the depths of tension failure from both analytical and numerical methods are almost the same for the studied slope.

3.3.2 Comparison of FOS

As previous description, the calculation of FOS is one of the most important aspects of seismic slope stability analysis. Table 3.2 shows the FOS from both analytical and numerical methods. From the Table 3.2, it can be found that the FOS estimated

Table 3.2 FOS calculated from various methods

Method		FOS	
		Tension-shear	Shear
Limit analysis	k_h = 0.5PGA/g = 0.1	1.0151	1.1001
	k_h = 0.75PGA/g = 0.15	0.9795	1.0261
FLAC^{3D}	Cut-through	0.98	1.11

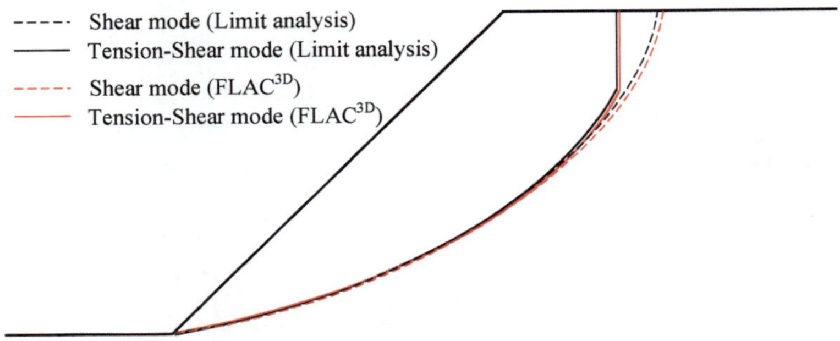

Fig. 3.21 Comparison of failure surface between limit analysis and FLAC^{3D} under two failure modes

from both limit analysis and FLAC^{3D} is almost the same for the studied slope. Noting again, the results from two different failure mechanisms can cause different judgments for the studied slope.

3.3.3 Comparison of the Shape of Slip Surface

Figure 3.21 shows the comparison of the failure surfaces from both limit analysis and FLAC^{3D} under two failure modes, from which almost the same failure surface under the same failure mode between these two methods can be found.

3.4 Conclusions

This chapter develops a new method for slope stability analysis based on both shear and tensile failure modes under seismic loadings. An analytical solution, limit analysis, of slope stability is derived based tension-shear failure mode. In addition, an approach is proposed for how to consider tension failure in addition to shear failure in widely used FLAC^{3D}, a finite difference method, so that stability analysis can be also carried out for a slope with complex slip surface. The purpose of this chapter is to investigate the influence of the depth of tension failure region on the FOS and the form of failure surface for slope stability analysis under earthquake loading. Based on the analyses above, the conclusions are drawn as follows:

(1) A larger number of analysis examples show that the influence of tension failure on slope stability is significant on FOS. FOS from tension-shear failure mode is smaller than that from traditional single shear failure mode. FOS could not be correct if tensile failure is ignored in the case of seismic loading.
(2) Results also show that tension failure has a significant influence on the failure surface of a slope. The shapes of failure surface of seismic slope under traditional single shear failure mechanism and tension-shear failure mechanism have an obvious difference. Failure surface of seismic slope considering tension-shear failure mechanism contains two segments and is shallower than that just considering the shear failure while ignoring the influence of tension failure.
(3) Comparing to the widely acceptable influence of horizontal seismic loading on slope stability, the vertical seismic loading should be taken more attention, especially for some slopes under extreme seismic loadings.
(4) The $FLAC^{3D}$ program can take the influence of tension failure into consideration, so as to make the analysis technique more reasonable for practical application. Results from a studied case show that $FLAC^{3D}$ gives almost the same FOS and failure surface compared to the limit analysis.
(5) Both limit analysis and $FLAC^{3D}$ can be used to determine reasonable FOS which is smaller than that obtained from traditional single shear failure mechanism. The wide use of this mechanism should now be seriously considered to traditional failure mechanism.

It should be noted that seismic slope stability analysis is a complex problem, especially for the seismic slope failure mechanism. The conclusions listed in here are based on limited observations. More in-depth and more extensive research should continue to be studied in future.

References

Antao, A. N., da Costa Guerra, N. M., Matos Fernandes, M., & Cardoso, A. S. (2008). Influence of tension cut-off on the stability of anchored concrete soldier-pile walls in clay. *Canadian Geotechnical Journal, 45*(7), 1036–1044.

Baker, R. (1981). Tenslie strength, tension cracks, and stability of slopes. *Soils and Foundations, 21*(2), 1–17.

Baker, R. (2003). Sufficient conditions for existence of physically significant solutions in limiting equilibrium slope stability analysis. *International Journal of Solids and Structures, 40*(13), 3717–3735.

Baker, R., & Leshchinsky, D. (2001). Spatial distribution of safety factors. *Journal of Geotechnical and Geoenvironmental Engineering, 127*(2), 135–145.

Chen, W.-F. (1975). *Limit analysis and soil plasticity*. Elsevier.

Chen, Y., & Xu, D. (2008). *FLAC/FLAC3D Foundation and engineering examples*. Beijing: Water Power Press of China.

Hassiotis, S., Chameau, J., & Gunaratne, M. (1997). Design method for stabilization of slopes with piles. *Journal of Geotechnical and Geoenvironmental Engineering, 123*(4), 314–323.

Huang, R. (2009). *Geohazard assessment of the Wenchuan earthquake*. Beijing: Science Press.

Huang, R., Xu, Q., & Huo, J. (2011). Mechanism and geo-mechanics models of landslides triggered by 5.12 Wenchuan earthquake. *Journal of Mountain Science, 8*(2), 200–210.

Itasca. (2007). *Fast Lagrangian Analysis of Continua in 3-Dimension (FLAC3D V3.1)*. Minnesota: Itasca Consulting Group.

Kaniraj, S. R., & Abdullah, H. (1993). Effect of berms and tension crack on the stability of embankments on soft soils. *Soils and Foundations, 33*(4), 99–107.

Lin, M., & Wang, K. (2006). Seismic slope behavior in a large-scale shaking table model test. *Engineering Geology, 86*(2–3), 118–133.

Spencer, E. (1968). Effect of tension on the stability of embankments. *Journal of Soil Mechanics & Foundations Div.*

Terzaghi, K. (1950). *Theoretical soil mechanics*.

Utili, S. (2013). Investigation by limit analysis on the stability of slopes with cracks. *Geotechnique, 63*(2), 140–154.

Wakai, A., Ugai, K., Sato, M., & Tazo, T. (2001). Numerical analysis for seismic behavior of a slope based on a simple cyclic loading model. In *4th International Conference on Recent Advances in Geotechnical Earthquake Engineering and Soil Dynamics, San Diego*.

Xu, Q., Pei, X., & Huang, R. (2009). *Large-scale landslides induced by the Wenchuan earthquake*. Beijing: Science Press.

Yin, Y., Wang, F., & Sun, P. (2009). Landslide hazards triggered by the 2008 Wenchuan earthquake, Sichuan, China. *Landslides, 6*(2), 139–152.

Zhang, Y., Chen, G., Zen, K., Kasama, K., & Dong, S. (2011a). Limit analysis of seismic slope stability based on tension-shear failure. In *Slope Stability 2011: International Symposium on Rock Slope Stability in Open Pit Mining and Civil Engineering, Vancouver, Canada*.

Zhang, Y., Chen, G., Zen, K., & Kasama, K. (2011b). High-speed starting mechanism of rock avalanches induced by earthquake. *International Conference on Advances in Geotechnical Engineering* (pp. 803–809). Australia: Perth.

Zhang, Y., Chen, G., Zen, K., Kasama, K., Wu, J., & Zheng, L. (2012a). Seismic slope stability analysis subjected to tension failure. In *Western Regional Division Report of Natural Disaster Research Council, Fukuoka, Japan* (pp. 21–24).

Zhang, Y., Chen, G., Wu, J., Zheng, L., & Zhuang, X. (2012b). Numerical simulation of seismic slope stability analysis based on tension-shear failure mechanism. *Geotechnical Engineering, 43*(2), 18–28.

Zheng, Y. R., Tang, X., Zhao, S., Deng, C., & Lei, W. (2009). Strength reduction and step-loading finite element approaches in geotechnical engineering. *Journal of Rock Mechanics and Geotechnical Engineering, 1*(1), 21–30.

Chapter 4
Run-Out Analysis of Earthquake-Induced Landslides

Abstract This chapter presents a long run-out model based on the so-called earthquake-induced trampoline effect and develops a practical numerical simulation program for estimating landslide movement behaviors. The multiple acceleration model (MAM) is derived from mechanism analysis of the earthquake-induced trampoline effect. The results show that the proposed new long run-out model is reasonable and applicable.

Keywords Landslides · Long run-out · Multiple acceleration model (MAM)

4.1 Introduction

Landslide is easily induced by a strong earthquake in mountainous areas and is potential serious threat to both lives and properties. It is important to estimate the run-out distance of landslide for disaster mitigation. In Japan, the *Law concerning disaster prevention due to landslide* has been promulgated and adopted by the government since 2003 (cited by Chen et al. 2010). The dangerous area for a potential landslide is identified as two times of slope height but less than 50 m in this law (Fig. 4.1). It is established based on statistics of the run-out distances in historical events, which were mainly triggered by heavy rains. The dangerous area identified by this law is not enough for earthquake-induced landslides.

Long run-out is one of the major movement behaviors of earthquake-induced landslides. Although some hypotheses have been proposed for explaining long run-out mechanism, they are helpless for explaining why so long distances occurred for the landslides induced by the 2008 Wenchuan earthquake since few of them consider the effect of seismic loading. However, with the reason of strong ground motion, landslides induced by earthquake can travel a distance several times longer

Y. Zhang, *Earthquake-Induced Landslides*, Springer Natural Hazards,
DOI 10.1007/978-981-10-2935-6_4

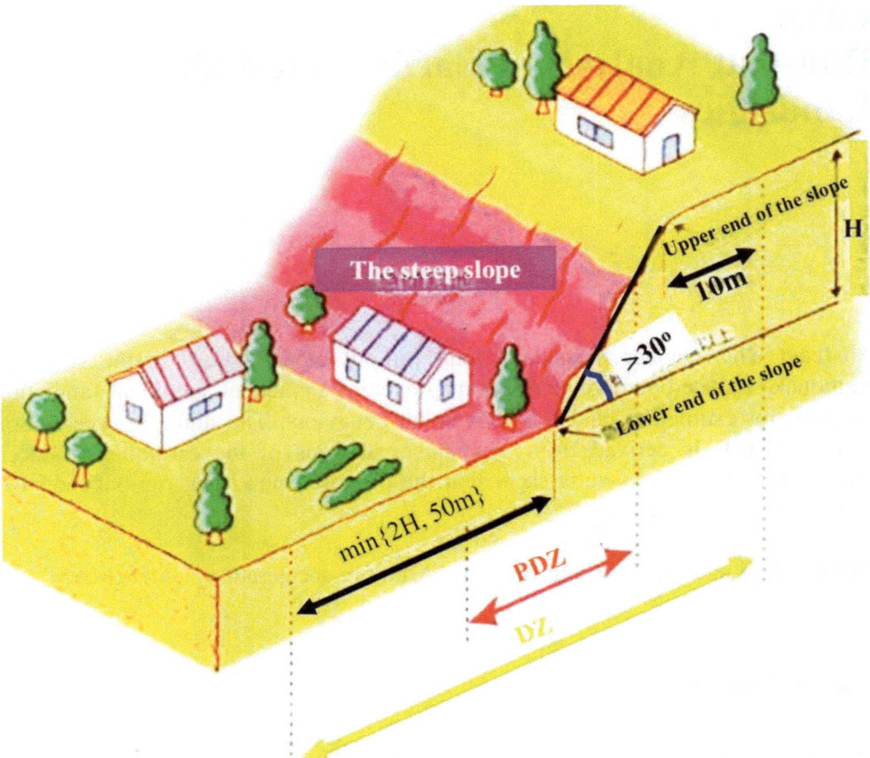

Fig. 4.1 Dangerous zone (*DZ*) for a prone landslide identified by law in Japan

than the height of the slope at a high velocity and result in great damages and losses (Keefer 1984, 2002; Rodriguez et al. 1999; Bird and Bommer 2004; Havenith and Bourdeau 2010). Some long run-out landslides induced by the 2008 Wenchuan earthquake are listed in Table 4.1. The case of Donghekou landslide induced by the 2008 Wenchuan earthquake shown in Fig. 4.2 is a classical long run-out landslide. The landslide destroyed all of the houses in the village which was located at the convergence of the Hongshi River and Xiasi River, and about 780 people were killed among about 1500 people who lived there. Figure 4.3 shows the photographs of the Donghekou area of pre- and after-landslide. The landslide has a height difference between the toe and main scarp of 700 m, a slide distance of 2400 m. The average thickness of the slide mass in the source area is about 80 m, with the maximum thickness of 110 m, and the total volume of the Donghekou landslide is

Table 4.1 Some long run-out landslides triggered by the Wenchuan earthquake (arranged from the horizontal distance)

No.	Name	Place	Area (m^2)	DF (km)	Volume (m^3)	Height (*H*/m)	Distance* (*D*/m)	*D*/*H*♀
1	Daguangbao	Anxian	7,273,719	4.8	7.5×10^{8}※–8.4×10^{8}◎	≈1450♣	4500※	3.10
2	Wenjiagou	Mianzhu	2,945,520	3.9	2.75×10^{7} -1.5×10^{8}※	1360	4170	3.07
3	Niujuangou	Wenchuan	527,700	0.3	7.5×10^{6}	827	2638–3102♠	3.31
4	Donghekou	Qiangchuan	1,283,627	0.3	1.0×10^{7}✲ -1.5×10^{7}	680	2413	3.55
5	Liqigou	Jiangyou	355,113	10.0	4×10^{6}	920	2310	2.51
6	Hongshigou	Anxian	687,520	2.2	2.6×10^{7}♥	≈900♣	2200♥	2.44
7	Shuimogou	Shifang	915,608	0.7	3.6×10^{7}♥	≈930♣	2100♥	2.26
8	Woqian	Qiangchuan	695,672	0.2	1.2×10^{7}	570	2043	3.58
9	Shibangou 1	Qiangchuan	496,983	2.3	7×10^{6}	710	1681♠–1829	2.58
10	Xiejiadianzi	Pengzhou	294,256	1.1	3.5×10^{6} -4.0×10^{6}✲	740	1500–1750♥	2.19
11	Chenjiaping	Anxian	169,368	1.1	1.2×10^{6}	680	1372	2.02
12	Baichaping	Dujaingyan	241,874	4.7	1.4×10^{6}	580	1340	2.30
13	Changtan	Mianzhu	151,094	6.7	1.4×10^{6}	800	1320	1.65
14	Zengjiashan	Mianzhu	198,165	11.4	2.2×10^{6}	700	1230	1.76
15	Xiaotianchi	Mianzhu	175,758	8.2	3.0×10^{6}	630	1120	1.78
16	Longwangou	Beichuan	99,821	0.7	540,382	520	890	1.71
17	Gaojiamo	Pingwu	115,301	1.6	1,126,823	340	722	2.12
18	Hongkouxiangxiajiaping	Dujaingyan	96,345	0.8	624,296	300	666	2.22
19	Beichuanzhongxuexinqu	Beichuan	124,365	0.3	2,400,000✲	≈300♣	664♠	2.21
20	Maochongshan 2	Pingwu	70,252	1.2	581,719	490	610	1.24
21	Wangjiayan	Beichuan	125,381	0.4	4,800,000✲	≈400♣	550✲	1.38

Note DF Distance to fault. * horizontal distance. ※data from Huang et al. (2012); ♥data from Wu et al. (2010); ♠data from Qi et al. (2011); ◎data from Chigira et al. (2010); and ✲data from Yin et al. (2009); ♣Estimated indirectly from geological sections or description; other data from Xu et al. (2009). ♀notes the value is mean one if the number of data more than one

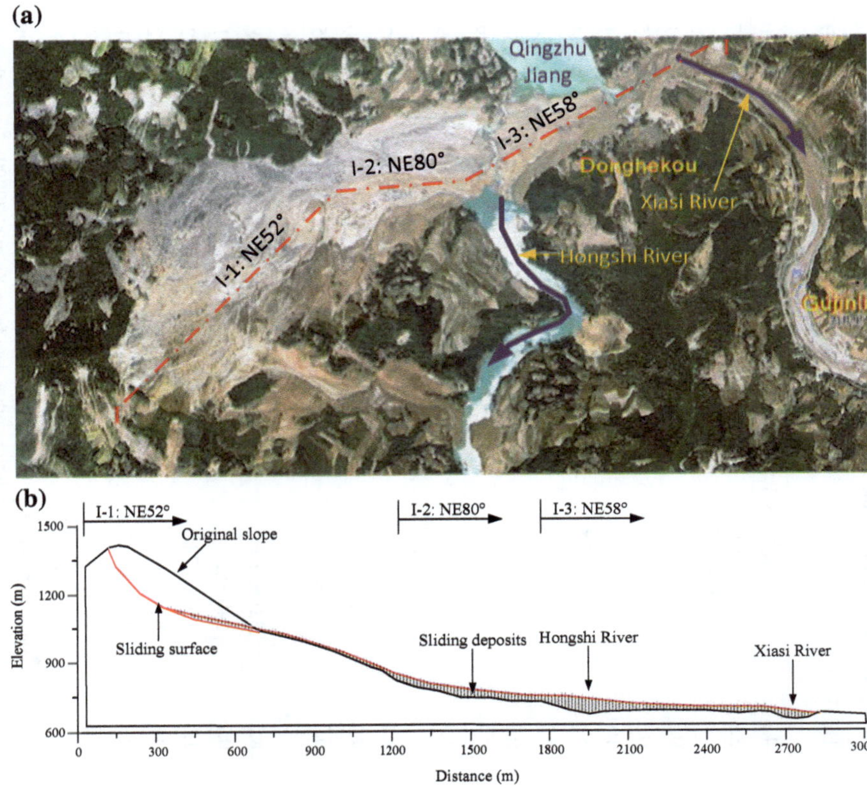

Fig. 4.2 Long run-out of earthquake-induced landslide. A typical long run-out landslide, Donghekou landslide, induced by the Wenchuan earthquake. **a** Air photo (modified from Yin et al. 2009) and **b** profit section of the Donghekou landslide

about 10–69 million m^3. The ratio of run-out distance to slope height is about 3.55, which is much larger than the value of general landslide, i.e., the dangerous area should be much larger for this kind of landslides.

Long run-out landslides listed in Table 4.1 are drawn in the map of earthquake zone with the seismogenic fault as shown in Fig. 4.4, from which close location to the fault of these long run-out landslides can be found. Therefore, the near-fault is found to be perhaps the reason of occurrence of the long run-out, and these phenomena must have somewhat relationship with the near-fault seismic loadings.

There are many unprecedented phenomena induced by the extreme near-fault seismic loadings appeared in recent earthquake events. Figure 4.5 shows an ejected

Fig. 4.3 Comparison of the Donghekou zone **a** before and **b** after earthquake. Four villages were destroyed and 780 people were killed by the Donghekou landslide

stone induced by the 2008 Wenchuan earthquake, which flied from the source area with a turning velocity and inserted into the ground surface. Also during the 2008 Wenchuan earthquake, a truck was found leaning against the wall after earthquake as shown in Fig. 4.6a. The similar jumped car was also found in the 2011 off the Pacific coast of Tohoku earthquake, as shown in Fig. 4.6b. It should be noted that

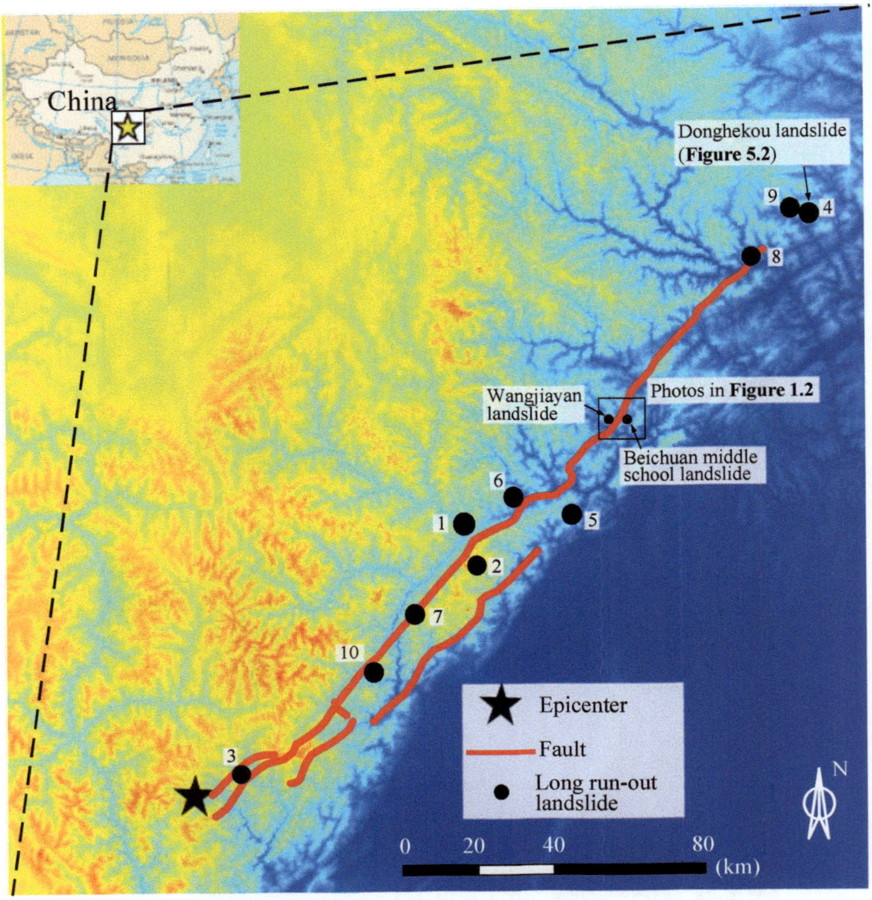

Fig. 4.4 Distribution of long run-out landslides induced by the 2008 Wenchuan earthquake. *Note* Number means the no. in Table 5.1

the extreme ground shaking can be occurred not only in the near-fault area, but also in the faraway-fault area if the earthquake magnitude is large enough, e.g., the site of threw up car in Fig. 4.6b is far away from the fault but the earthquake magnitude is M9.

Consequently, this study attempts to evaluate the effects of extreme seismic loadings on the landslide run-out by a new run-out model for earthquake-induced landslides and simulates the long run-out of landslides by using extended DDA program.

Fig. 4.5 Rejected stone induced by the extreme near-fault seismic loadings from the 2008 Wenchuan earthquake

(a)

(b)

Fig. 4.6 Two jumped cars induced by the 2008 Wenchuan earthquake and the 2011 off the Pacific coast of Tohoku earthquake, respectively (*photograph* **a** is from Tang)

4.2 Trampoline Effect Induced by Extreme Seismic Loadings

4.2.1 Extreme Seismic Loading

An extreme earthquake wave was recorded during the June 14, 2008, Iwate-Miyagi earthquake (Ms. 6.9) in Japan. The vertical component of peak accelerations reaches as larger as 4 g recorded at the West Ichinoseki station, 3 km southwest of the epicenter. Moreover, the vertical component is more than twice its horizontal counterpart for this extreme ground motion, which is exactly the opposite of the general knowledge that horizontal ground motion is much larger than vertical motion.

Incomprehensible thing is that the phenomenon was not confirmed or supported by the downhole [(ground level) GL–260 m] sensor set at the same station where only the peak accelerations of 683 and 1036 gal were recorded for the vertical and horizontal directions, respectively (Fig. 4.7).

Some people were even sceptical about the unprecedented record. However, the instruments and records have been carefully checked and no problem has been found. Aoi et al. (2008) proposed a possible mechanism for this extreme movement using the concept of *trampoline effect*.

They suppose that soil and rocks near-surface may lose their cohesion through the development of tensile cracks and apertures, and the near-surface layer of the ground behaves like an open box filled with sand. When there is upward acceleration, the sand, subject to compressional stress, behaves elastically, and this may continue to be true under moderate downward acceleration. As the downward acceleration increases, however, the sand particles may begin to lose mutual contact and fall into a virtual free-fall state.

They give an analogy between the movement and that of an athlete bouncing on a trampoline by the following approaches: (a) A distinct polarity asymmetry waveform (Fig. 4.8a) is assumed to represent the motion of an undeformable mass bouncing on a trampoline by cyclic oscillations, (b) a selected part of the downhole record is used to represent the elastic deformation of a deformable mass (Fig. 4.8b, c), and (c) the motion of a deformable mass bouncing on a trampoline is then obtained as the sum of the above two waveforms (Fig. 4.8c).

Actually, Aoi et al. (2008) successfully demonstrated the waveform asymmetry that the upward accelerations reached as large as about 5 g, while the downward accelerations were bounded near $-1 \times g$. However, we notice that there is no explanation of why and how the distinct polarity asymmetry waveform as shown in Fig. 4.8a comes into being, especially, the upward acceleration of as large as 4000 gal. Also, what does the box represent in their model of an open box filled with sand.

If the near-surface layer behaves like sand, it is obvious that the ground below the near-surface layer should take the role of the box. This implies that the lower ground should have 4000 gal upward acceleration to compress the sand and more

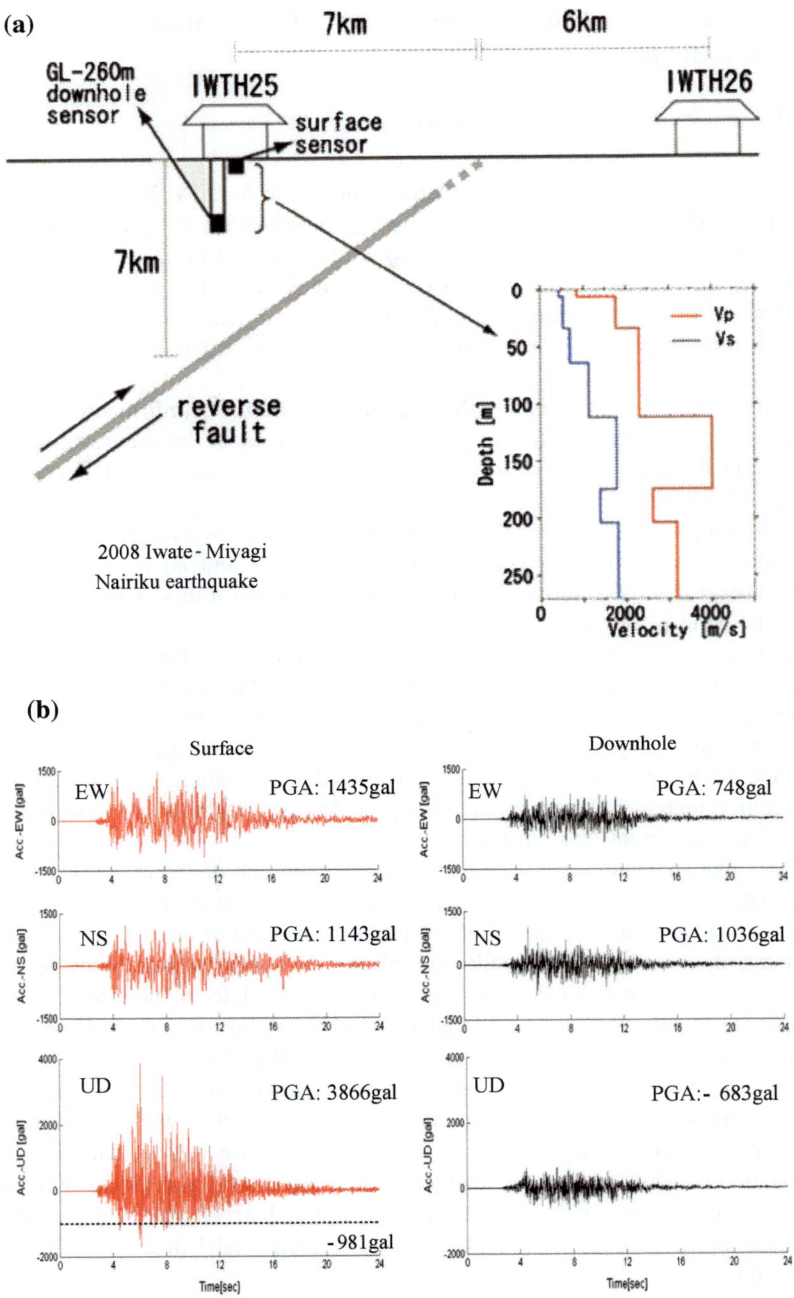

Fig. 4.7 West Ichinoseki station and its waveforms (after Aoi et al. 2008) **a** fault and sensor locations; **b** acceleration waveforms for both surface and downhole sensors

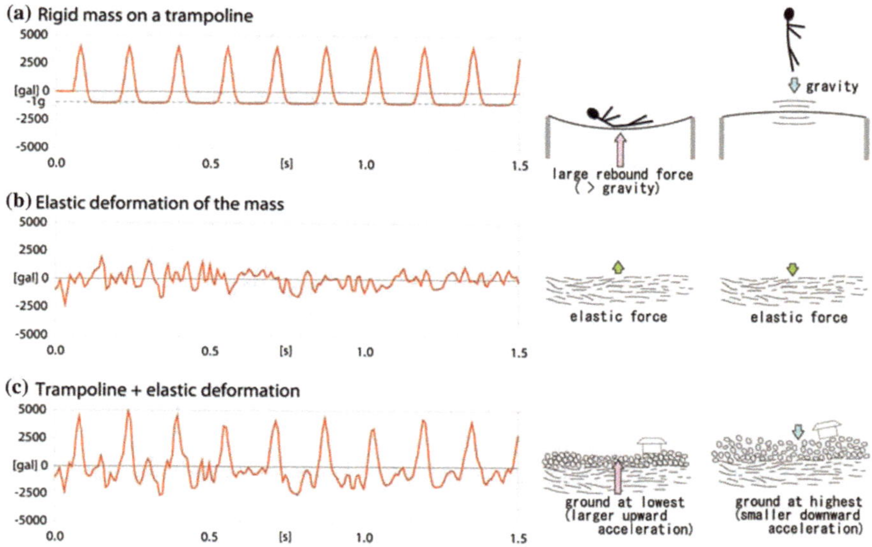

Fig. 4.8 **a** Simplified model of the motion of an undeformable mass bouncing on a trampoline. **b** Elastic deformation of a deformable mass, represented by a selected part of a downhole seismic record. **c** Simulated motion of a deformable mass bouncing on a trampoline, obtained as the sum of (**a**) and (**b**) (after Aoi et al. 2008)

than 1 g downward accelerations to let the sand free-fall. Unfortunately, the downhole record does not support the requirement since the recorded waveform is almost symmetric and its maximum amplitude is less than 700 gal.

For this reason, this study gives an explanation why and how the trampoline movement as shown in Fig. 4.8a comes into being. By using DDA numerical simulation, we can reproduce an asymmetric waveform with large upward accelerations and less than 1 g downward accelerations.

4.2.2 Trampoline Effect

As we know, while an athlete is bouncing on a trampoline, the potential energy from the height, the kinetic energy and the potential energy from the springs of the trampoline are interchangable. Neglecting energy loss in the energy transformation, the jump height would keep unchanged if there is neither additional energy nor total energy loss involved. In order to jump higher and higher, the athlete can obtain additional energy by propelling his body up into the air using his flexed leg muscles. On the other hand, the athlete can loss the energy by changing his posture touching the mat.

While an object is bouncing on the ground, although the same energy transformation as trampoline may occur, it cannot obtain additional energy by itself in ordinary situation. However, an object can obtain additional energy from the vibrating ground during a strong earthquake since the object is impacted by the vibrating ground just like a ball is hit by a racket. Also, it is possible for the object to lose its energy through the vibrating ground as follows.

Considering a single wave, we divide one period into two phases: P-phase and N-phase. The upward part of the period is called P-phase, and the downward part is called N-phase. If the object is touching the ground during the P-phase, it could get additional energy from the impact of the ground motion. On the other hand, if the object is touching the ground during the N-phase, it could lose the energy due to the longer contact time because the ground is moving in the same direction.

It is important to notice that the object can be obtained as large as two times the velocity and several times the acceleration of the ground from the impact of the vibrating ground. We show this based on the elastic collision theory just by considering the fact that the mass of the ground is much larger than the mass of the object as follows.

By way of simplification, consider two blocks (Fig. 4.9). Let M and m be the masses, v_{in} and V the velocities before collision, and V' and v_{out} the velocities after collision, respectively. According to the principles of the conservation of both energy and momentum, we can obtain the velocity of small block after collision.

$$\begin{aligned} MV - mv_{\mathrm{in}} &= MV' + mv_{\mathrm{out}} \\ MV^2 + mv_{\mathrm{in}}^2 &= MV'^2 + mv_{\mathrm{out}}^2 \end{aligned} \tag{4.1}$$

$$v_{\mathrm{out}} = \frac{2V + (1 - m/M)v_{\mathrm{in}}}{1 + m/M} \tag{4.2}$$

If terms $k_{\mathrm{vpr}} = \frac{1-m/M}{1+m/M}$ and $k_{\mathrm{vtr}} = \frac{2}{1+m/M}$ are introduced, then Eq. 5.2 can be translated as

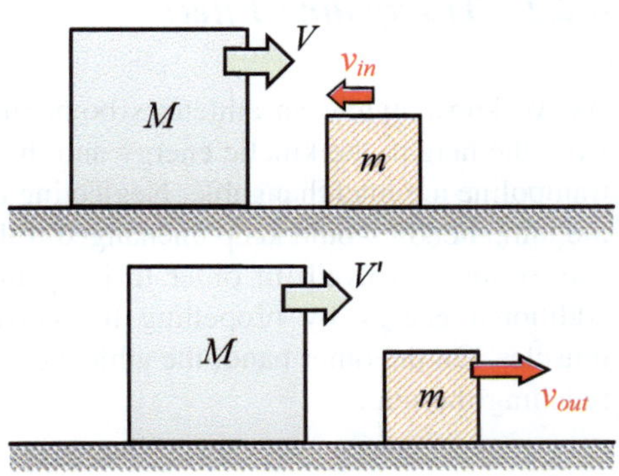

Fig. 4.9 Collision model

$$v_{out} = k_{vtr}V + k_{vpr}v_{in} \quad (4.3)$$

where k_{vpr} is called the velocity preserve rate with values of (−1 to 1), and k_{vtr} is velocity transmittable ratio with values of (0–2).

When the ground, considered as a huge block M, impacts an object m on the ground, k_{vpr} becomes 1 and k_{vtr} becomes 2, since the mass of the ground M is so huge that the ratio of m to M can be neglected. Thus, the velocity of the object after collision can be estimated as follows:

$$v_{out} = 2V + v_{in} \quad (4.4)$$

It can be seen that the object can obtain an additional velocity of as large as two times that of the ground from the collision, which behaves as a racket ball hit by a racket bat. Thus, an object on the ground can get higher and higher due to a sequence of collisions caused by earthquake, which behaves as an athlete bouncing on a trampoline. We call this kind of movement as trampoline motion.

We also find that the acceleration of an object can obtain more than two times that of the ground because of the so-called trampoline effect. If an object is bouncing off the ground due to the ground impaction, the upward acceleration would be larger than 1 g, and its downward acceleration would be 1 g because of free-fall motion back to the ground. Therefore, it is obvious that the ground is not necessary to have acceleration more than 1 g in order to make an object on it obtain the trampoline motion.

4.2.3 *Verification of Trampoline Effect Using DDA*

The collision between a large block and a small block is analyzed by DDA. The large block with mass of M has the initial velocity of 10 m/s, and the small block with mass of m is at a standstill. The two blocks have the same size $1m \times 1m$ and are located on a fixed rectangular block. The cohesion, friction, and tension strength are set to zero between blocks. The physical parameters and control parameters are shown in Table 4.2. The large block is simulated by changing its density.

Table 4.2 Physical parameters and control parameters of collision model used in DDA simulation

Parameter	Value
Density (Kg/m^3)	2000
Young's modulus (Pa)	10^9
Poisson' ratio	0.1
Penalty	10^8
Time step (s)	0.001

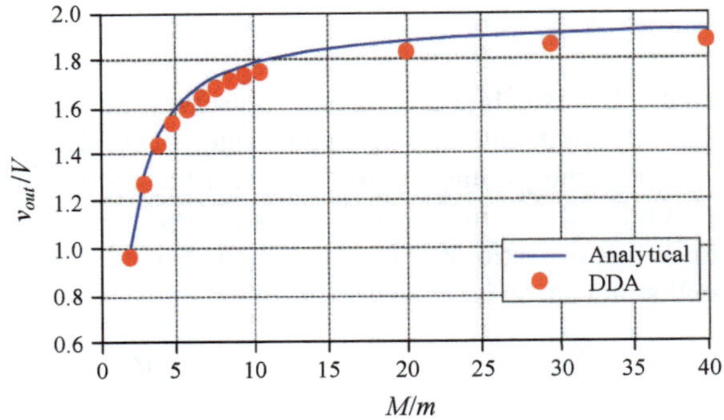

Fig. 4.10 v_{out}/V from analytical result and DDA result

After the large block impacted the small block, the velocities of both blocks changed. The results obtained from DDA simulations by changing M are shown in Fig. 4.10, together with the theoretical analytical values, in which v_{out}/V is equal to the k_{vtr} in Eq. 4.3 ($v_{in} = 0$) for theoretical values (the line) and calculated by the ratio of v_{out}/V in DDA simulation (the dots).

It can be seen that the k_{vtr} obtained from DDA is in quite good agreement with the analytical solution. However, by close examination, it can be found that the k_{vtr} values from DDA are little smaller than the analytical values when the ratio of M to m is larger than four. This is because elastic strains of the two blocks are led to energy transformed into potential energy of deformation by the collision in DDA simulation, while no strain is considered in analytical solution.

Furthermore, if the small block has an initial velocity toward the large one, the v_{out}/V could become larger and larger. It can be considered as the trampoline effect caused by a series of vibration.

4.3 Existing Long Run-Out Models

The extraordinarily long run-out of landslide was apparently first recognized by Heim (cited by Hsü 1975), who stated that the deposit of the Elm event, and the descriptions of eyewitnesses, suggested the characteristics of a fluid flow rather than of a sliding accumulation of solids. And it has drawn more and more attention worldwide since Vaiont rockslide of 1963 which fell into a reservoir with disastrous results (Davies 1982).

The existing models to interpret the long run-out mechanism can be broadly divided into four categories as described in the following sections:

4.3.1 Fluidizations

Heim rationalized this behavior by suggesting that the internal motion of the debris consisted of a myriad of highly energetic collisions among individual grains, the result of which was to tend to maintain the original kinetic energy of the fall.

Hsü (1975) developed the grain-contact idea of Helm by linking it with the grain-flow theory of Bagnold (1954).

McSaveney (1978) suggested that due to the vibrational energy imparted by the earthquake, the debris mass became dilated and fluidized and could slide a long run-out distance, at the same time thinning and spreading.

Erismann (1979) proposed that a thin layer of molten rock between flow and slope might facilitate the motion, by friction decreasing and high pressure increasing.

A hypothesis of acoustic fluidization was proposed by Melosh (1979). The mechanism described that a high-frequency vibration might be capable to relieve the static overburden pressure in limited regions of the rock mass, allowing sliding to occur.

Foda (1994) and Kobayshi (1997) investigated from the wave-theoretical point of view. That a decaying wave guided along the basal layer could catch up with the slide mass in high slide velocities exceeding a threshold. This causes a 'sonic boom' phenomenon that may contribute to the loosening of sliding masses. The velocities on long slopes are more likely to exceed the threshold value.

4.3.2 Air Cushion

Kent (1966) proposed a totally different mechanism that the debris masses were maintained in a dilated state by the rapid upward flow of air escaping from within the body of the debris. Such a dilatation would cause friction decreasing that allows the debris to travel for long distances.

Shreve (1968) also invoked air as the mobilizing agent. He suggested that the debris mass, having jumped over a height or impacted into the ground, could trap and compress a layer of air beneath it. The high-pressure air could reduce sliding friction by partly supporting the sliding overburden.

Goguel (1978) discussed the air pressure of steam, generated by frictional heat between landslide mass and slope, also could reduce the sliding friction.

Yue (2013) proposed ground gas eruption mechanism, a new genesis and mechanism responsible for causing large, rapid, and long run-out landslide. He pointed that during an earthquake, a huge amount of natural gas with high pressure quickly escapes the deep crust traps and migrates and flows through tectonic faults, geological faults, weak zones, and/or porous channels into the rock masses forming mountains or hillside slopes. Since the rock materials can have relatively high strengths and rigidity, the rock mass can resist the initial loading of the quickly migrate. The erupted gas, however, can seep into the rock masses and fully saturate their voids and fissures and apertures. Eventually, the expanding powers of the gas in the rock masses would instantly break, erupt, and/or explode the rock masses with limited confining in situ pressures and force them to flow out of their original positions and form the landslides with broken rock masses and long distance ejections and flows.

4.3.3 Liquefaction

Sassa (1996) conducted undrained cyclic loading ring shear tests and proposed a concept of sliding surface liquefaction as a key mechanism of rapid and long travel landslides. Sassa (2000) succeeded in simulating the triggering process by applying record of real seismic wave. The main cause of this phenomenon is the grain crushing along the sliding surface which let the soil skeleton volume tends to shrink, and it leads to excess pore pressure which reduces effective stress and shear resistance.

In the models mentioned above, the vibration of slope in earthquake is only considered as an initial mechanism and the energy transferred from earthquake to landslide mass is not considered to generate rapid velocity and long run-out distance. For this point, a new long run-out model is presented and a practical numerical simulation method is developed to validate the new model in Sects. 5.5 and 5.6.

4.4 Multiplex Acceleration Model (MAM) and Extension

In order to analyze the mechanism of high speed and long run-out phenomenon of earthquake-induced landslide, our group have proposed a multiple acceleration model (MAM). The MAM considers two cases of landslide with and without earthquake. The vibration of a slope caused by earthquake wave can be divided into two phases: P-phase and N-phase. When the slope moves in the outer normal direction of the slope surface, it is called P-phase wave. The debris on the surface will be pushed and accelerated by the slope in this phase. When the slope moves in the inner normal direction of the slope surface, it is called N-phase wave. Since slope surface moves apart from the debris, the friction should be declined (Chen et al. 2013).

The MAM can be used to explain some long run-out landslide induced by earthquake. However, the role of the vertical seismic loading is not clear and the trampoline effects are mainly induced by the vertical seismic loading. In order to consider the trampoline effect of earthquake loading on the landslide run-out process, the MAM is extended in the next subsection.

This section extends the original MAM by introducing movement translation. There are at least four types of movement for a collapse mass from a slope, sliding, falling, rolling, and collision, in which the sliding is the most general movement type. A single sliding movement for general landslide is presented firstly, and the effects of earthquake loadings on the movement translation are discussed by considering trampoline effect.

4.4.1 Single Sliding Model

Supposing that sliding block with mass m moves from position A to position B during a landslide, the potential energy decreases by mgh. Based on the energy conservation law, it is easy to obtain the following equation for a sliding stone movement in the general situation (case 1 in Fig. 4.11).

$$mgh_s - \sum_{i=1}^{n} l_i mg \tan \varphi_{si} \cos \theta_i = 0 \tag{4.5}$$

The first term here is for potential energy and the second term is for the work of sliding friction force between the slope and the falling block, where only the general sliding movement is considered and the whole curve path is divided into finite linear segments. And m = mass, g = gravity acceleration, h_s = the sliding height, l = the segment length, θ is the segment slope angle, $\tan \varphi_s$ is the sliding friction coefficient, and i is the index of segment. Obviously, run-out distance $D_s = \sum_{i=1}^{n} l_i \cos \theta_i$.

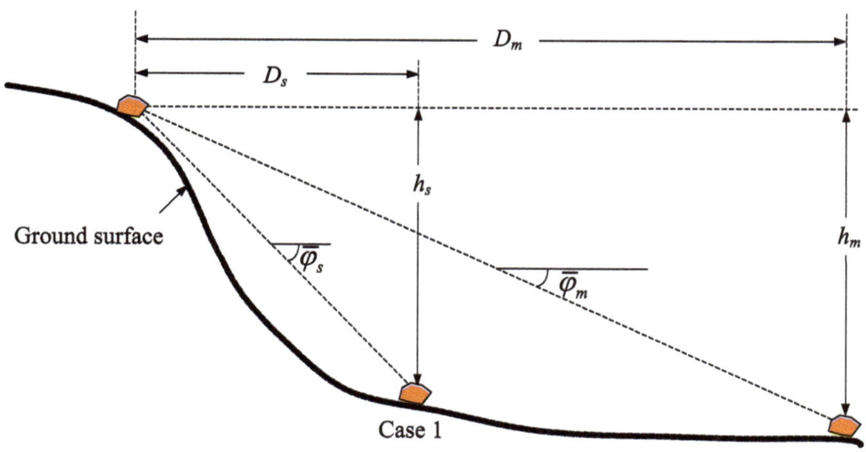

Fig. 4.11 Comparison of run-out distance in terms of apparent friction angle from two cases: *Case 1* single sliding model and *Case 2* multi-movement model

The angle of reach (Fig. 2.7 in Sect. 2.3.2), usually used for the discussion of run-out distance can be obtained from Eq. 4.5 as follows:

$$\tan \bar{\varphi}_{\mathrm{s}} = \frac{h_{\mathrm{s}}}{D_{\mathrm{s}}} = \sum_{i=1}^{n} w_i \tan \varphi_{\mathrm{s}i} \tag{4.6}$$

where $w_i = D_{\mathrm{s}i}/D_{\mathrm{s}}$. Specially, if the friction angle for every segment is the same, the apparent friction angle $\bar{\varphi}_{\mathrm{s}}$ will equal to the friction angle φ_{s}.

When we consider the effects of movement translation, three translations, (i) sliding translate to falling, (ii) sliding translate to rolling, and (iii) sliding translate to collision, are considered (case 2 in Fig. 4.11).

Thus, Eq. 4.5 becomes

$$\begin{aligned} & mgh_m - \underbrace{\sum_{i=1}^{n} l_i mg \cos\theta_i \tan\varphi_{\mathrm{s}i} -}_{\text{Sliding}} \\ & \underbrace{\sum_{j=1}^{m} l_j mg \cos\theta_j \tan\varphi_{\mathrm{r}j}}_{\text{Rolling}} - \underbrace{\sum_{p=1}^{M} l_p mg \cos\theta_p \mu_f}_{\text{Falling}} - \underbrace{\sum_{k=1}^{N} \frac{1}{2} m\left(1-\eta_k^2\right)\left(v_k^{\mathrm{pre}}\right)^2}_{\text{Collision}} = 0 \end{aligned} \tag{4.7}$$

where h_m is the multi-movement height. The whole run-out distance is divided into three parts for three movements of sliding, rolling, and falling. The third term here is for the work of rolling friction force between the slope and the collapse block in the rolling movement stage, in which j is the index of total m segments of rolling and $\tan\,\varphi_r$ is the rolling friction coefficient. The forth term is for the energy dissipation during the falling period, in which μ_f is the falling friction coefficient and p is the index of the total M times of falling. The fifth item in Eq. 4.7 is for the kinetic energy of a falling stone obtained from the collision with the vibrating slope, in which v_k^{res} is the residual velocity in kth collision and η_k is the ratio of residual velocity to precollision velocity, $\eta_k = \frac{v_k^{\mathrm{out}}}{v_k^{\mathrm{in}}}$, usually η_k is the rebound coefficient k_{re} between the collapse block and the ground and $0 < \eta_k = k_{\mathrm{re}} < 1$ in static situation.

In the multi-movement model, the run-out distance D_m can be expressed as sum of distance from three types of movement:

$$D_m = D_{\mathrm{s}} + D_r + D_f = \underbrace{\sum_{i=1}^{n} l_i \cos\theta_i}_{\text{Sliding}} + \underbrace{\sum_{j=1}^{m} l_j \cos\theta_j}_{\text{Rolling}} + \underbrace{\sum_{p=1}^{M} l_p \cos\theta_p}_{\text{Falling}} \tag{4.8}$$

It should be noted that the energy dissipation from the falling movement is usually zero, but it often along with a collision in which energy dissipation may occurred. If we consider the energy dissipation from the falling movement is zero, i.e., the falling friction coefficient $\mu_f = 0$, and the rolling friction coefficient tan φ_r is discount of the sliding friction coefficient tan φ_s, the apparent friction angle for the case 2 in Fig. 4.11 can be obtained from Eq. 4.7 as follows:

$$\tan \bar{\varphi}_m = \frac{h_m}{D_m} = \sum_{i=1}^{m+n} w_i k_i^* \tan \varphi_{si} + \frac{1}{2gD_m} \sum_{k=1}^{N} \left(1 - \eta_k^2\right) \left(v_k^{\text{pre}}\right)^2 \tag{4.9}$$

where k_i^* is the multiple coefficient of conveying from sliding to rolling friction coefficient and from static to dynamic friction coefficient, usually $k_i^* < 1$.

Comparing Eq. 4.9 with Eq. 4.6, it can be seen that the apparent angle $\bar{\varphi}_m$ from the multi-movement model can be smaller or larger than that from the single sliding model. The existing long run-out model, e.g., fluidization, air cushion, and liquefaction, can be considered as these effects reduce the multiple coefficient k_i^* in Eq. 4.9.

4.4.2 *Multi-movement Model*

When we take the seismic loading into account to the movement translation model (Eq. 4.9), two effects may be occurred: (1) k_i^* in Eq. 4.9 may be smaller than that in static situation, since the ground shaking may aggravate the discount of friction coefficient; and (2) the ration of residual velocity to precollision velocity η_k may be larger than unit due to the ground shaking, i.e., the residual velocity may be larger than the precollision value. In another word, the collapse block can obtain the energy from the ground shaking. These two effects are shown in Fig. 4.12. This study mainly focuses on the collision effect by considering trampoline effects.

Figure 4.13 shows the different rebounding velocity of a collapse block from two phases of the ground movement. Different bounding velocities can be obtained from a simply energy balance expression, which are given as Eq. 4.10. The first term in the right hand of Eq. 4.10 notes the bounding velocity from an unshaking ground, i.e., static situation. The second term notes the changed velocity because of the ground movement. From Eq. 4.10, we can find that the bounding velocities are reduced and increased from N-phase and P-phase, respectively. However, the velocity of P-phase is often much larger than that of N-phase because of the trampoline effects. Hence, the energy dissipation of block from the N-phase is smaller than the obtained energy of block from the P-phase. Taking into account Eq. 4.10, the bound coefficient for two phases can be obtained from Eq. 4.11. The bound coefficient η_k in Eq. 4.9 would have different signs from these two phases. Also, as the upward acceleration may be much larger than downward one under the trampoline effects, the increased velocity in P-phase would be much larger than that

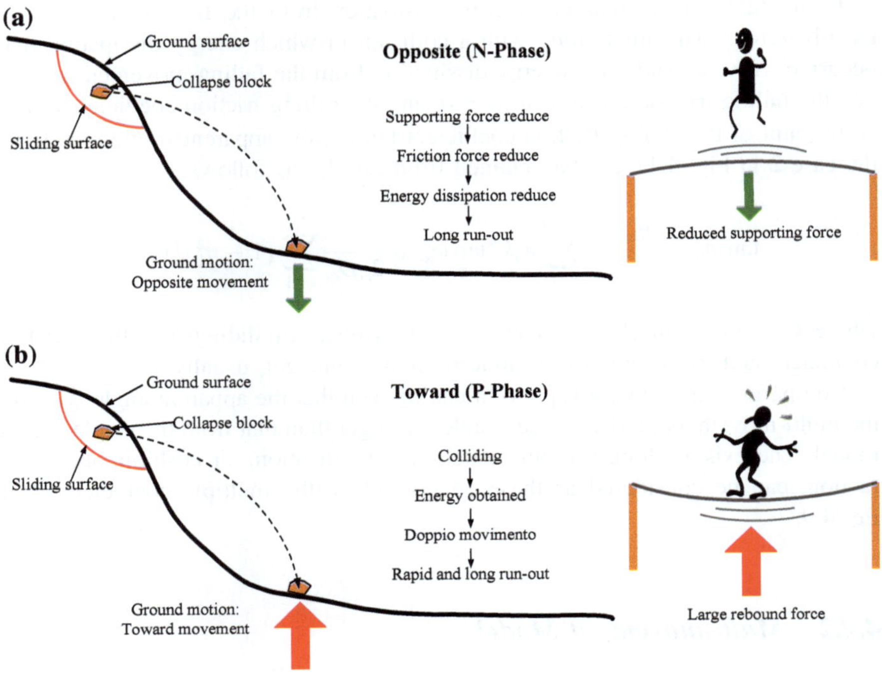

Fig. 4.12 Trampoline effect of earthquake-induced landslides

in N-phase. In a ground shaking, the velocity from these two phases can be obtained from Eq. 4.12.

$$\begin{aligned} v_{out\_N} &= k_{re} v_{in} - k_{vtr} V_N \\ v_{out\_P} &= k_{re} v_{in} + k_{vtr} V_P \end{aligned} \tag{4.10}$$

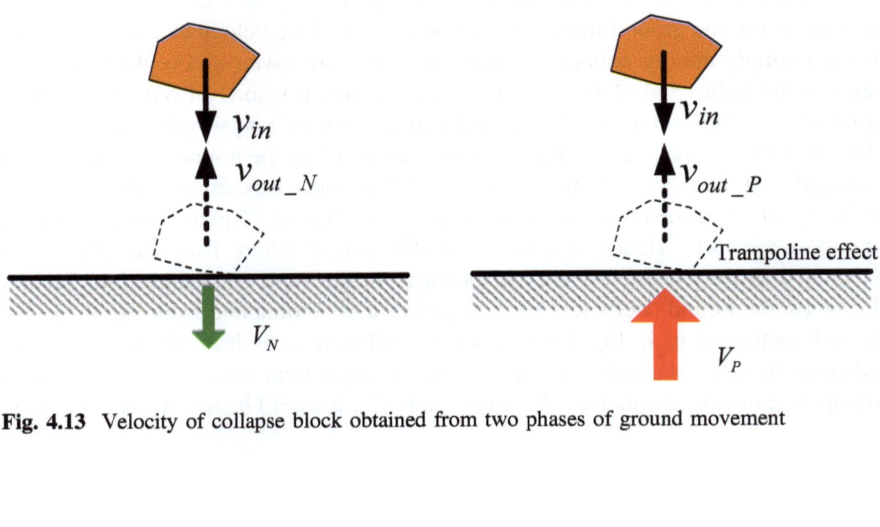

Fig. 4.13 Velocity of collapse block obtained from two phases of ground movement

$$\eta_N = k_{\text{re}} - k_{\text{vtr}} \frac{V_N}{v_{\text{in}}}$$
$$\eta_P = k_{\text{re}} + k_{\text{vtr}} \frac{V_P}{v_{\text{in}}} \tag{4.11}$$

$$V_N = \int_t^{t+\Delta t} a_{(-)}(t)dt$$
$$V_P = \int_t^{t+\Delta t} a_{(+)}(t)dt \tag{4.12}$$

$a_{(-)}(t)$ and $a_{(+)}(t)$ are the accelerations of slope vibration due to earthquake. Because of the trampoline effect of the extreme seismic loading, the velocity from P-phase will be larger and larger. Figure 4.13 shows the bounding velocity from different P-phase velocity, in which DDA results are also given.

In order to take the MAM account into the landslide run-out analysis, a widely used DDA program is used to incorporate and validate the MAM and simulate the long run-out of earthquake-induced landslide in the next chapter (Fig. 4.14).

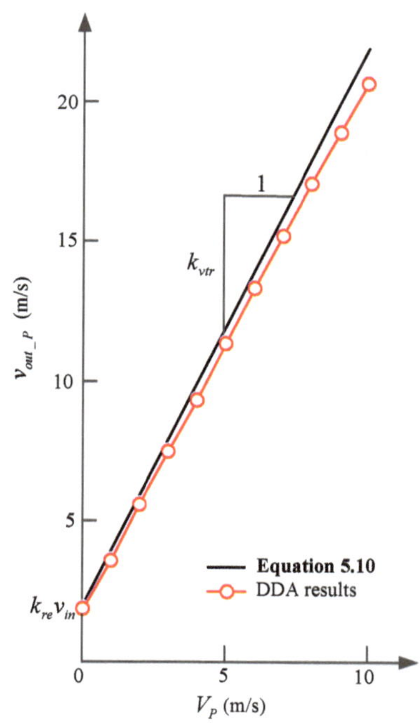

Fig. 4.14 $v_{\text{out\_}P}$ against the velocity of P-phase

References

Aoi, S., Kunugi, T., & Fujiwara, H. (2008). Trampoline effect in extreme ground motion. *Science, 332*(5902), 727–730.

Bagnold, R. A. (1954). Experiments on a gravity-free dispersion of large solid spheres in a Newtonian fluid under shear. *Proceedings of the Royal Society of London. Series A, 225*(1160), 49–63.

Bird, J. F., & Bommer, J. J. (2004). Earthquake losses due to ground failure. *Engineering Geology, 75*(2), 147–179.

Chen, G., Zheng, L., & Zhang, Y. (2013). Practical applications of DDA to disaster prevention. In *ICADD 11*. NRC, Fukuoka, Japan

Chen, G., Zen, K., Zheng, L., & Jiang, Z. (2010). A new model for long-distance movement of earthquake induced landslide. In *The 44th U.S. Symposium on Rock Mechanics*, Salt Lake City, UT, USA

Chigira, M., Wu, X., Inokuchi, T., & Wang, G. (2010). Landslides induced by the 2008 Wenchuan earthquake, Sichuan China. *Geomorphology, 118*(3–4), 225–238.

Davies, T. R. (1982). Spreading of rock avalanche debris by mechanical fluidization. *Rock Mechanics, 15*(1), 9–24.

Erismann, T. (1979). Mechanisms of large landslides. *Rock Mechanics, 12*(1), 15–46.

Foda, M. (1994). Landslides riding on basal pressure waves. *Continuum Mechanics and Thermodyn, 6*(1), 61–79.

Goguel, J. (1978). Scale-dependent rockslide mechanisms, with emphasis on the role of pore fluid vaporization. *Rockslides and avalanches, 1,* 693–705.

Havenith, H. B., & Bourdeau, C. (2010). Earthquake-induced landslide hazards in mountain regions. A review of case histories from central asia. *Geologica Belgica, 13*(3), 137–152.

Hsü, K. J. (1975). Catastrophic debris streams (sturzstroms) generated by rockfalls. *Geological Society of America Bulletin, 86*(1), 129–140.

Huang, R., Pei, X., Fan, X., Zhang, W., Li, S., & Li, B. (2012). The characteristics and failure mechanism of the largest landslide triggered by the Wenchuan earthquake, May 12, 2008 China. *Landslides, 9*(1), 131–142.

Keefer, D. K. (1984). Landslides caused by earthquakes. *Geological Society of America Bulletin, 95*(4), 406.

Keefer, D. K. (2002). Investigating landslides caused by earthquakes—a historical review. *Surveys in Geophysics, 23,* 473–510.

Kent, P. (1966). The transport mechanism in catastrophic rock falls. *The Journal of Geology, 74* (1), 79–83

Kobayashi, Y. (1997). Long runout landslides riding on basal guided wave. *Engineering Geology and the environment, Balkema, Rotterdam* (pp. 761–766).

McSaveney, M. (1978). Sherman glacier rock avalanche, Alaska, USA. *Rockslides and avalanches, 1,* 197–258.

Melosh, H. J. (1979). Acoustic fluidization: A new geologic process?. *Journal of Geophysical Research: Solid Earth* (1978–2012), 84(B13), 7513–7520

Qi, S., Xu, Q., Zhang, B., Zhou, Y., Lan, H., & Li, L. (2011). Source characteristics of long runout rock avalanches triggered by the 2008 Wenchuan earthquake. *China. Journal of Asian Earth Sciences, 40*(4), 896–906.

Rodriguez, C. E., Bommer, J., & Chandler, R. J. (1999). Earthquake-induced landslides 1980–1997. *Soil Dynamics and Earthquake Engineering, 18*(5), 325–346

Sassa, K. (1996). Prediction of earthquake induced landslides. In *Proceedings of 7th International Symposium on Landslides, Trondheim," Landslides"*,Vol.1, pp. 115–132

Sassa, K. (2000). Mechanism of flows in granular soils. In *Proceedings of the International Conference of Geotechnical and Geological Engineering*, GEOENG2000, Melbourne, Vol.1, pp. 1671–1702,

Shreve, R. L. (1968). Leakage and fluidization in air-layer lubricated avalanches. *Geological Society of America Bulletin, 79*(5), 653–658.

Wu, S., Wang, T., Shi, L., Sun, P., Shi, J., Li, B., et al. (2010). Study on catastrophic landslides triggered by 2008 great Wenchuan earthquake, Sichuan China. *Journal of Engineering Geology, 18*(2), 145–159.

Xu, Q., Pei, X., & Huang, R. (2009). *Large-scale landslides induced by the Wenchuan earthquake*. Beijing: Science Press.

Yin, Y., Wang, F., & Sun, P. (2009). Landslide hazards triggered by the 2008 Wenchuan earthquake, Sichuan China. *Landslides, 6*(2), 139–152.

Yue, Q. Z. (2013). Natural gas eruption mechanism for earthquake landslides: Illustrated with comparison between Donghekou and Papandayan rockslide-Debris flows. In *Earthquake-Induced Landslides* (pp. 485–494). Springer

Chapter 5
Extension of Discontinuous Deformation Analysis and Application in Run-Out Analysis of Earthquake-Induced Landslides

Abstract This chapter extends the original discontinuous deformation analysis (DDA). A practical numerical simulation program is developed by incorporating the MAM into the extended DDA. After an extreme ground movement with the peak ground acceleration (PGA) of 4000 gal is successfully reproduced, some large-scale landslides induced by the 2008 Sichuan Earthquake are analyzed in practical numerical simulations. The results show that the movement behaviors of earthquake-induced landslides can be analyzed using the numerical simulation program.

Keywords Discontinuous deformation analysis (DDA) · Earthquake-induced landslides · Extreme ground movement

5.1 Introduction

Discontinuous deformation analysis (DDA) (Shi 1988; Shi and Goodman 1985) is a discrete numerical method that was developed for computing large deformation and large displacement in discontinuous block system. DDA introduces a unified format for the consideration of not only the translation, rotation, and deformation of an individual block but also such movement forms as sliding and opening along block boundaries, having the advantages of both the distinct element method (DEM) and the finite element method (FEM) (Jiang and Yeung 2004). Since the novel formulation and the numerical code of DDA were presented, DDA draws more and more attention and many modifications and improvements to the original formulas have been proposed to overcome some of its limitations (Ke 1996; Lin et al. 1996; Chen et al. 1997; Koo and Chern 1998; Cheng 1998; Doolin and Sitar 2004; Doolin 2005; Wang et al. 2013) and make it more efficient, suitable, and practical to engineering computations, e.g., rockfall (Chen 2003; Wu et al. 2005; Ma et al. 2011; Chen et al. 2013), landslide (Sitar and Maclaughlin 1997; Sitar et al. 2005; Wu 2004, 2007, 2010; Kveldsvik et al. 2008; Wu et al. 2009; Wu and Chen 2011; Zhang et al. 2012, 2013), tunnel (Yeung and Leong 1997; Tsesarsky and Hatzor 2006), blast (Ning et al. 2010, 2011a, b), dynamic block (Hatzor and Feintuch

Y. Zhang, *Earthquake-Induced Landslides*, Springer Natural Hazards,
DOI 10.1007/978-981-10-2935-6_5

2001; Maclaughlin et al. 2001; Doolin and Sitar 2002; Tsesarsky et al. 2005; Bakun-Mazor et al. 2012), and others (Yeung 1993; Hatzor and Benary 1997; Kim et al. 1999; Thomas and Bray 1999; Kamai and Hatzor 2008). In addition, a number of high-profile projects were studied by the DDA, e.g., the Three Gorges Dam project in China (Dong et al. 1996; Zhu et al. 1999), Pueblo Dam in Colorado (Kottenstette 1999), the Yerba Buena tunnel portal, San Francisco (Law and Lam 2003), Norway's Gjovik Olympic Cavern (Scheldt et al. 2002), and Israel's Masada National Monument (Hatzor et al. 2002, 2004).

The behavior of a block system depends on the strength of block boundaries, including cohesion c, inter-friction angle φ (or friction coefficient $\tan\varphi$), and tension strength σ_t. In the original DDA code, the Mohr–Coulomb failure criterion is applied in block interface and the penalty technical is used to prevent interpenetration between blocks. In order to validate the accuracy of DDA, many researchers carried out comparisons of results between DDA and other methods, including (i) analytical methods, (ii) other numerical methods, and (iii) laboratory and field investigation. In these validation cases, a single block on an inclined plane is the simplest and the most efficient one. If the strength of interface is only with friction, the errors are generally lower than 1% (MacLaughlin et al. 2001; Doolin and Sitar 2002; Tsesarsky et al. 2005; Kamai and Hatzor 2008). However, once the strength of joint contains cohesion, the accuracy will be reduced sharply; i.e., the original DDA cannot simulate the failure behavior of cohesive material accurately. That's why even one validation example for cohesive material does not exist in more than 100 validation cases referenced by a review paper (MacLaughlin and Doolin 2006).

In order to find the reason of DDA cannot accurately simulate failure of cohesive material, some researchers tried their best in several ways. Accurate calculation of failure depends on the calculation of contact force in interface. Many researchers take the opinion that the calculation of contact force in DDA is impossible to provide an accurate result because the penalty technique an accurate result due to the penalty technique is used and stiffness of penalty spring is artificial. Because the original DDA often shows an inordinate stability compared to real situation for cohesive material, strength degradation method is often used to improve the accuracy. The strength of interfaces can depend on displacement (Sitar et al. 2005; Wang et al. 2013) or velocity (Tsesarsky et al. 2005; Bakun-Mazor et al. 2012). The displacement- or velocity-dependent shear strength can improve the accuracy, but the functions c (*displacement* or *velocity*) and φ (*displacement* or *velocity*) are often defined by the user optionally. So, the failure behavior of cohesive material in DDA is still an unsolved problem.

The purpose of this paper is to present an extended DDA that can accurately simulate the failure behavior of cohesive-frictional material. In original DDA, one edge-to-edge contact is treated as two vertex-to-edge contacts that may have the different contact states, and once one of the vertex-to-edge contacts failed, half of cohesion between the joint is improperly removed. This unreasonable situation, one joint has two contact states along with two treatments of cohesion, is the reason of the original DDA cannot simulate the cohesive material accurately. In this study, an additional evaluation of edge-to-edge contact state is added into the original DDA code to avoid the situation of two contact states exist in one edge-to-edge contact. A series of analytical solutions and DDA results are presented to validate the accuracy of the extended DDA.

5.2 Theory of DDA

5.2.1 Basic Theory and Time Discretization

The discontinuous deformation analysis (DDA) is formulated based on block theory where each block can move and deform independently, and the interaction between blocks is idealized by contact springs. For a two-dimensional problem, each block of arbitrary geometry has six degrees of freedom, among which three components are rigid body motion terms and the other three are constant strain terms. So the deformation variable of block i can be written as follows:

$$D_i = \left(u_0 v_0 r_0 \varepsilon_x \varepsilon_y \gamma_{xy}\right)^{\mathrm{T}} \tag{5.1}$$

where u_0 and v_0 are the translations of block centroid (x_0, y_0) along the x- and y-axes, r_0 is the rigid rotation around (x_0, y_0), and $\left(\varepsilon_x \varepsilon_y \gamma_{xy}\right)$ are the normal and shear strains of block at (x_0, y_0). The displacement $U = (u, v)$ at any point (x, y) of a block can be represented as follows:

$$U = TD_i \tag{5.2}$$

where the displacement transformation matrix T is defined as follows:

$$T = \begin{bmatrix} 1 & 0 & -(y-y_0) & x-x_0 & 0 & (y-y_0)/2 \\ 0 & 1 & x-x_0 & 0 & y-y_0 & (x-x_0)/2 \end{bmatrix} \tag{5.3}$$

DDA computation takes the first-order approximation of the displacement function, which represents the constant stress and strain at any arbitrary point within the block. The rock is assumed to be elastic, and the shear resistance at the boundary is assumed to follow the Mohr–Coulomb yield criterion when the rocks make contact with each other.

A system of blocks is formed from the individual blocks through contacts among blocks and displacement constraints on individual blocks. For the block system, the simultaneous equilibrium equations are similar to the system equations in the FEM, i.e., Hamilton's principle and minimized potential energy.

$$M\ddot{D} + C\dot{D} + KD = F \tag{5.4}$$

where D, $\dot{D}, \ddot{D}$ are the matrixes of displacement, velocity, and acceleration, respectively; M is the mass matrix, C is the damping matrix, K is the stiffness matrix, and F is the forcing matrix. The damping matrix C in Eq. 5.4 can be rewritten as follows in terms of viscosity η and mass matrix M:

$$C = \eta M \tag{5.5}$$

The physical meaning of viscosity η is the damping of the rock itself, the viscosity of air around the rock surfaces, and the vegetation on the surface of a rock slope. In this study, no viscous damping is introduced and the energy loss is Coulomb in nature due to the adoption of Mohr–Coulomb yield criterion that controls the block sliding.

The kinematic Eq. 5.4 is solved by Newmark's β and γ method by using parameters $\beta = 0.5$ and $\gamma = 1.0$ as Eqs. 5.6 and 5.7:

$$D_{n+1} = D_n + \Delta t_n \dot{D}_n + \frac{\Delta t_n^2[(1-2\beta)\ddot{D}_n + 2\beta\ddot{D}_{n+1}]}{2} \tag{5.6}$$

$$\dot{D}_{n+1} = \dot{D}_n + \Delta t_n[(1-\gamma)\ddot{D}_n + \gamma\ddot{D}_{n+1}] \tag{5.7}$$

where subscript n notes calculation step 'n' in the computations. From Eqs. 5.6 and 5.7, assuming the initial displacement D_n at calculation step 'n' is 0 because the updating Lagrange descriptions are used in the analysis. Then

$$\ddot{D}_{n+1} = \frac{D_{n+1} - D_n - \Delta t_n \dot{D}_n - \Delta t_n^2(1-2\beta)\ddot{D}_n/2}{\beta \Delta t_n^2} \tag{5.8}$$

Substitution of Eq. 5.8 into Eq. 5.2 forms the global form.

The algebraic equation for the increase in displacement is solved for each time increment by the following equation:

$$\widetilde{K} \cdot D = \widetilde{F} \tag{5.9}$$

where $\widetilde{K}$ and $\widetilde{F}$ are effective stiffness matrix and effective force matrix, respectively. Assuming that a block system consists of n blocks, we have

$$\widetilde{K} = \begin{bmatrix} K_{11} & K_{12} & \ldots & K_{1n} \\ K_{21} & K_{22} & \ldots & K_{2n} \\ \ldots & \ldots & \ldots & \ldots \\ K_{n1} & K_{n2} & \ldots & K_{nn} \end{bmatrix}, D = \begin{bmatrix} D_1 \\ D_2 \\ \ldots \\ D_n \end{bmatrix} \text{ and } \widetilde{F} = \begin{bmatrix} F_1 \\ F_2 \\ \ldots \\ F_n \end{bmatrix} \tag{5.10}$$

where D_i, F_i (i = 1, 2, …, n) are 6 × 1 submatrices, D_i is the deformation variable of block i, F_i is the load distributed to the six degrees of freedom of block i, K_{ij} (i, j = 1, 2, …, n) is a 6 × 6 submatrices, K_{ii} is relevant to the material properties of block i, and K_{ij} is defined by the contact between blocks i and j.

For the dynamic landslide simulation in this chapter, the ground excitation applying on base block is employed as acceleration input and forms the force matrix F in Eq. 5.4. In addition, this dynamic numerical method gives unique solutions for problems having large displacement and deformation by evaluating correct contact patterns between rocks during a landslide. The large displacement behavior of each block can be traced by updating the coordinates of each block at the end of each calculation step.

The flowchart of the procedures of the DDA program is shown in Fig. 5.1.

5.2.2 Contact Mechanism

Blocks in a block system can contact with each other at their boundaries. For a two-dimensional problem, there are three possible contact types, vertex-to-vertex, vertex-to-edge, and edge-to-edge (Fig. 5.2). In the original DDA, an edge-to-edge contact is treated as two vertex-to-edge contacts. For example, the edge-to-edge contact of P_1P_2 to P_3P_4 shown in Fig. 5.2c can be transformed into two vertex-to-edge contacts of P_2 to P_3P_4 and P_3 to P_1P_2. Hence, only two contact types, vertex-to-edge and vertex-to-vertex, are calculated in the original DDA code. This process reduces the computational cost and simplifies the program.

Penalty technique is used in the original DDA to prevent interpenetration between blocks. In original DDA code, when two blocks contact with each other and some penetration occurred, deformed spring(s) can be added in normal and/or sliding direction(s) to top out the penetration. The degree of spring deformation depends on the depth of penetration. In fact, at each time step, add or minus spring (open-close iteration) is used to enforce no-tension, no-penetration conditions for every contact before proceeding to the next time step. Figure 5.3 shows the contact process of DDA in a step time. If Mohr–Coulomb failure criterion is used in DDA, contact state can be determined by the following criteria.

$$k_n d_n > 0 \text{ and } k_s d_s \leq k_n d_n \tan\varphi + cl \rightarrow \textit{locked} \tag{5.11}$$

$$k_n d_n > 0 \text{ and } k_s d_s > k_n d_n \tan\varphi + cl \rightarrow \textit{sliding} \tag{5.12}$$

$$k_n d_n \leq 0 \rightarrow \textit{open} \tag{5.13}$$

where k_n and k_s are stiffness of the added normal and shear springs, respectively (Fig. 5.3b), and they can be defined by the user according to the properties of the material; generally, k_n is set as 10–1000 times of elastic modulus of material and k_s is set as a quarter of k_n, i.e., $k_s = k_n/4$; d_n and d_s are the distances of penetration in normal and shear directions, respectively (Fig. 5.3b); φ is the joint inter-friction angle; c is the joint cohesion; and l is the calculated contact length, specially, if the vertex-to-edge is one of vertex-to-edge contacts of an edge-to-edge contact, l is half of contact length of the edge-to-edge contact.

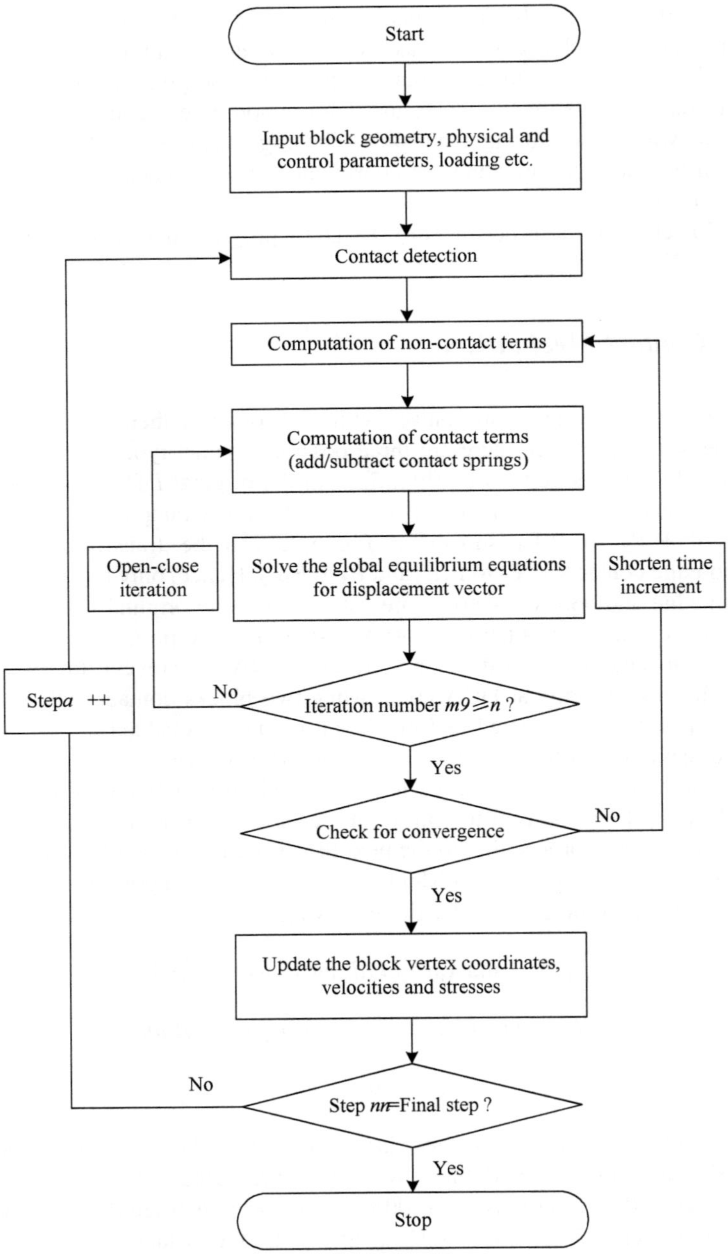

Note: n is defined by the user, default as 6

Fig. 5.1 Flowchart of the procedures of the DDA program

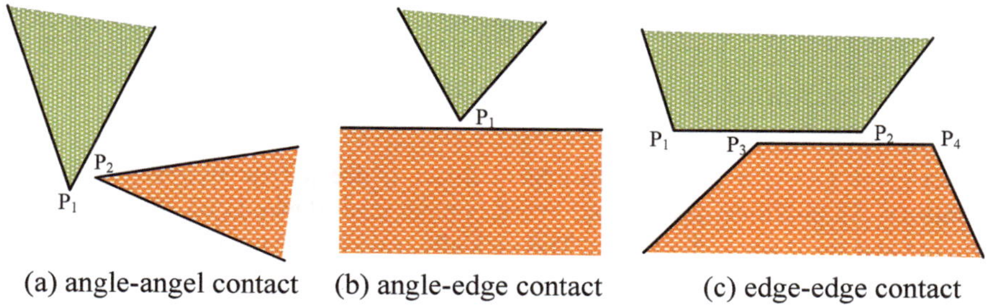

Fig. 5.2 Three possible contact types in 2D block analysis: **a** vertex-to-vertex contact, **b** vertex-to-edge contact, and **c** edge-to-edge contact

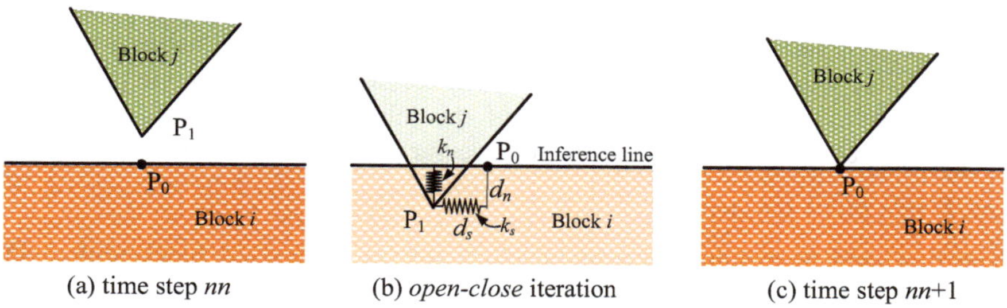

Fig. 5.3 A vertex-to-edge contact process of DDA in the time step of nn to $nn + 1$

The contact states locked, sliding, and open refer to the kind of penalty constraint that applies to a contact. Figure 5.4 shows the three states by using a vertex-to-edge contact. For locked contact state (Fig. 5.4a), both normal and shear springs are added into the equilibrium equations, and normal spring and shear friction are added for sliding contact state (Fig. 5.4b). For open contact state (Fig. 5.4c), neither spring nor friction is added into the equilibrium equations. It should be noted that the cohesion will be removed once a contact state change from locked to sliding or open in the original DDA. This assumption is appropriate for analyzing jointed material-like rock systems.

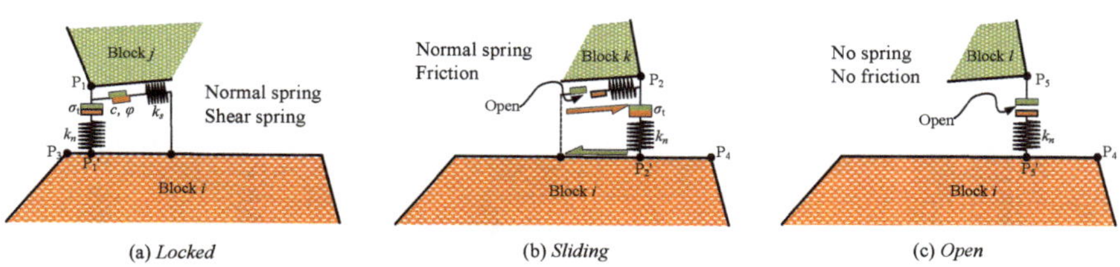

Fig. 5.4 Three states of a vertex-to-edge contact

5.3 Validation of Static Sliding

5.3.1 Limitation of the Original DDA

In original DDA, every edge-to-edge contact is treated as two vertex-to-edge contacts. As previous description, every contact state depends on the penetration degree and shear strength of joint. Theoretically, two vertex-to-edge contacts of one joint should have the same contact state. However, they often show different contact states due to different penetration degree, especially when shear strength is near to the critical value that makes the limit contact state. Thus, in some time step, two contact states may exist in one joint; e.g., sliding vertex-to-edge contact (Fig. 5.4a) and locked vertex-to-edge contact (Fig. 5.4b) may exist in one edge-to-edge contact (Fig. 5.5). In addition, the cohesion of sliding vertex-to-edge contact is removed; i.e., parts of cohesion between the joint are improperly removed that accelerate the failure of the locked vertex-to-edge contact in the same joint. Then, the whole joint is destroyed. The developing process of the joint is shown in Fig. 5.6. Some examples are carried out to examine the accuracy of the process in the following.

(a) *Single block on incline*

The first kind of example is a single block resting on a fixed block with a 45° incline (Fig. 5.7). The resting block is allowed to slide alone the incline and has three shapes for three examples as shown in Fig. 5.7a–c. The geometrical parameters for three examples are shown in Fig. 5.7. Three resting blocks have the same contact length of 0.57 m with the fixed block. The resting blocks in examples 2 and 3 have two times area of the block in example 1. In order to describe the contact state of the two blocks, the number of vertices is shown in schematic model. Physical parameters of the blocks and control parameters are listed in Table 5.1. To obtain acceptable displacements and stresses, the contact spring stiffness of 100 times of elasticity modulus is used in here. The full dynamic analysis is used to make the velocity of block to inherit the one in last time step. Theoretically, as long as rolling is not occur, the sliding blocks in examples 2 and 3 need the same shear strength to reach the critical stability state and the shear strength is twice of that in example 1. In the examining examples, both cohesion and friction are considered.

Since the full dynamic analysis is used, once the block is unstable, the displacement of sliding block is increasing and the contact state between the sliding block and fixed block is sliding. Otherwise, the displacement is keep at zero and the

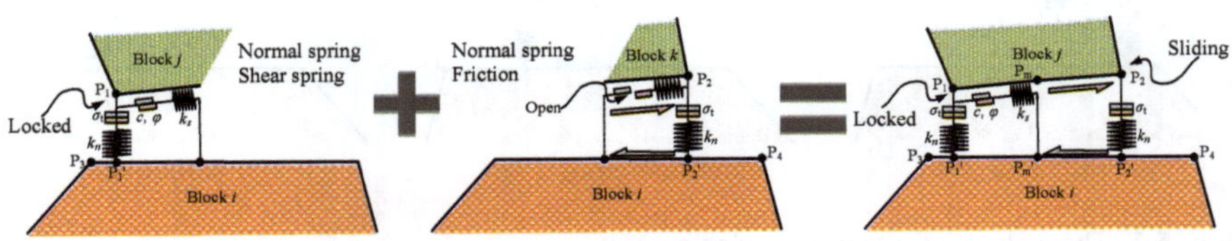

Fig. 5.5 A possible contact state in the original DDA

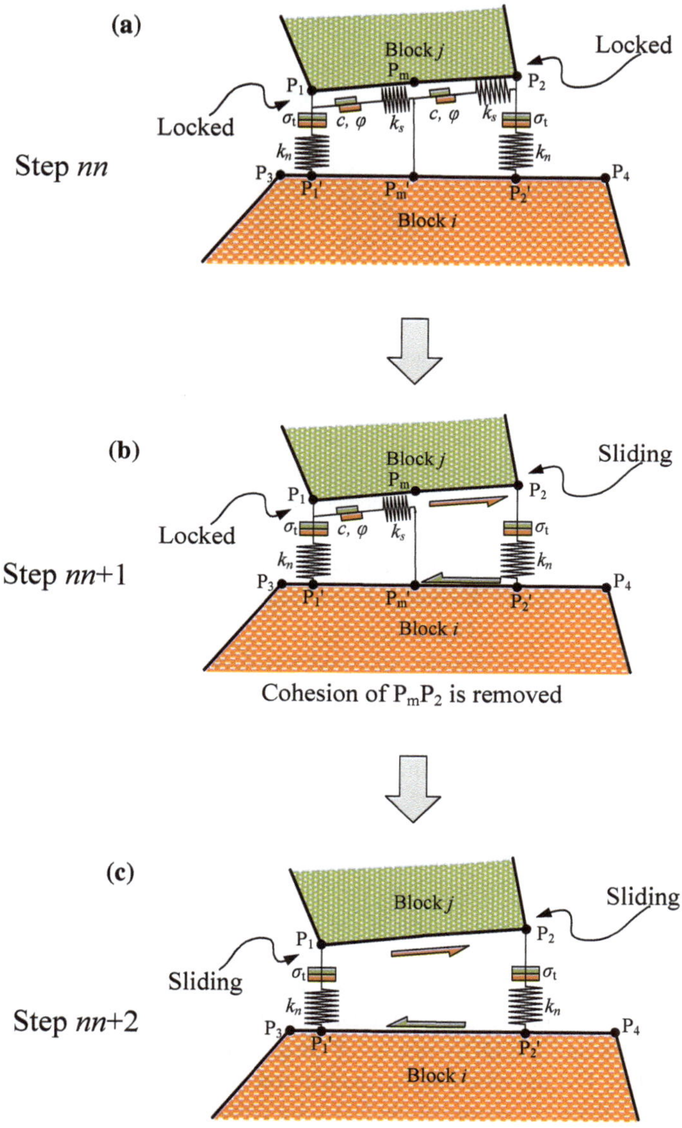

Fig. 5.6 A possible developing process of edge-to-edge contact in the original DDA

contact state is locked. To obtain the correct critical shear strength, it is essential to trace the cohesion gradually until the contact state is changed from sliding to locked. When the cohesion is no smaller than the critical value c^{critical}, the sliding block will be locked to the fixed block and the displacement will be kept at zero, whereas once the cohesion is larger than the critical value, the sliding block will be

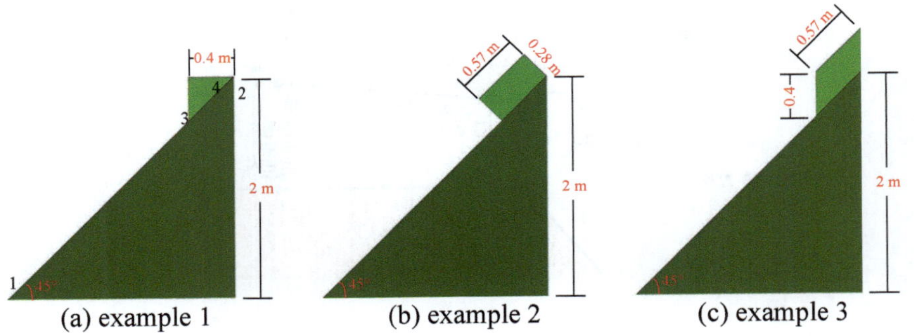

Fig. 5.7 Three examples for single block on incline

Table 5.1 Parameters for examining examples

Parameter	Value
Unit weight, γ (kN/m^3)	20
Elasticity modulus, E (kPa)	1.0 × 106
Poisson's ratio, v	0.2
Dynamic control parameter	1
Maximum displacement ratio, g_2	0.001
Time step, g_1 (s)	0.01
Contact spring stiffness, k_n (kN/m)	1.0×10^8
Factor of over-relaxation	1.0

slide along the incline surface and the displacement will be increased. For the example of Fig. 5.7a with $\varphi = 30°$, a critical cohesion of 1.10 kPa is obtained after a series of trial calculation, and when $c < c^{\text{critical}}$ the sliding block will be unstable and when $c \geq c^{\text{critical}}$, the sliding block will be stable. The judge criterion of critical stability is shown in Fig. 5.8.

Figure 5.9 shows the comparison of the critical stability analysis results between the original DDA and the analytical method, in which different cohesions are obtained for different given friction angles based on the judge criterion presented in the previous paragraph. In Fig. 5.9, the DDA results are shown to be larger than the analytical results and the DDA results of cases 2 and 3 are shown to be different from each other. For the cases in the shaded area in Fig. 5.9, the results should be stable, but the calculated results of the DDA are unstable. The results from DDA are even unreasonable for some situations. In Fig. 5.9, the pairs of DDA results line and the corresponding analytical solution divided the cases with different combination of cohesion and friction angle into three parts: (i) consistent stable cases in the upper of the DDA results line (upper cases), within which the sliding block should theoretically be stable and the DDA results consist with the analytical solution; (ii) inconsistent cases between the DDA results line and analytical solution (shaded cases), within which the sliding block should theoretically be stable but the DDA regards the block as unstable; and (iii) consistent cases in the lower of the

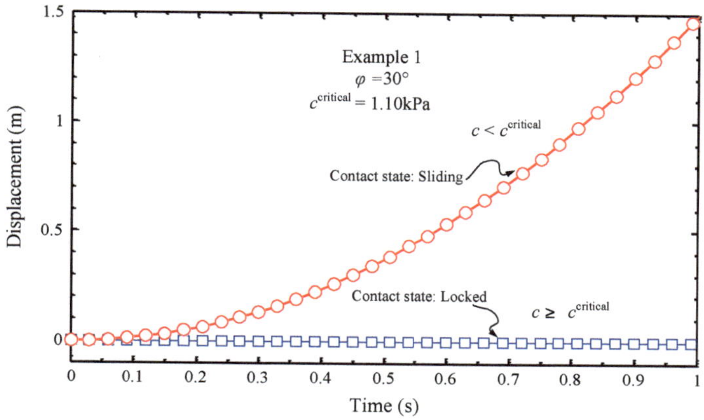

Fig. 5.8 Judge criterion of critical stability

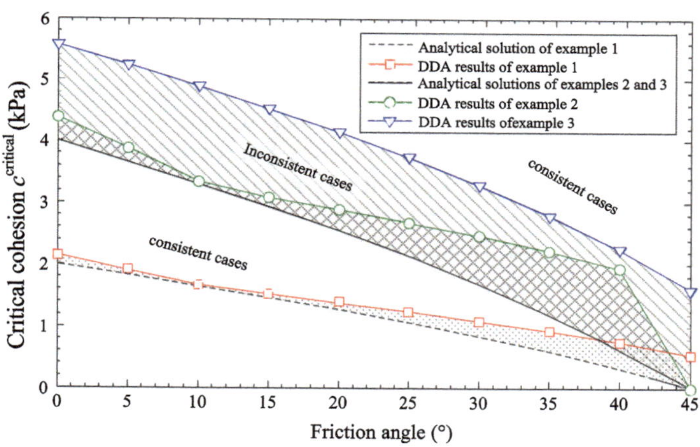

Fig. 5.9 Critical stability analysis results of three examples for single block on incline

analytical solution (lower cases), within which the sliding block should be theoretically unstable and the DDA results consist with the analytical solution. In addition, the relative errors of the shear strengths required for the critical stability calculated by DDA are shown in Fig. 5.10. The relative errors, E_{rel}, is defined as follows:

$$E_{\text{rel}} = (F - 1) \times 100\% = (\frac{cl + W \cos\theta \tan\varphi}{W \sin\theta} - 1) \times 100\% \tag{5.14}$$

where l is the contact length, θ is the slope angle, and W is weight of sliding block. The relative errors for examples 1 and 2 vary within the range between 0 and 35%, whereas it is about 40% for example 3.

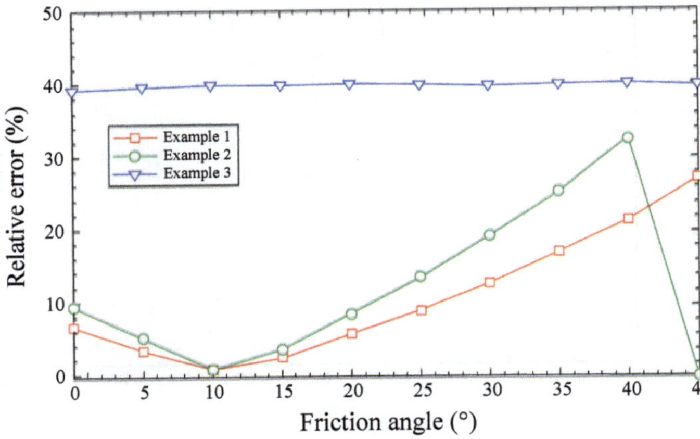

Fig. 5.10 Relative errors of the shear strength required for the critical stability analysis calculated by DDA

In order to find the reason of these inconsistent cases in the shade area of Fig. 5.9, contact states for each contact at each time step are recorded, in which sliding state recorded as number 1 and locked state recorded as number 2. Figure 5.11 shows the DDA movement results and contact states of five cases: two consistent cases (Fig. 5.11a, b) and three inconsistent cases (Fig. 5.11c–e). As expect, for the consistent cases, both the movement result and the contact state are reasonable. However, for those inconsistent cases, the movement results are unreasonable, e.g., the cases in Fig. 5.11d, e, in which almost the same displacements are observed for two cases with large difference of inter-friction angle. In addition, the unreasonable developing process of edge-to-edge contact (Fig. 5.6) is

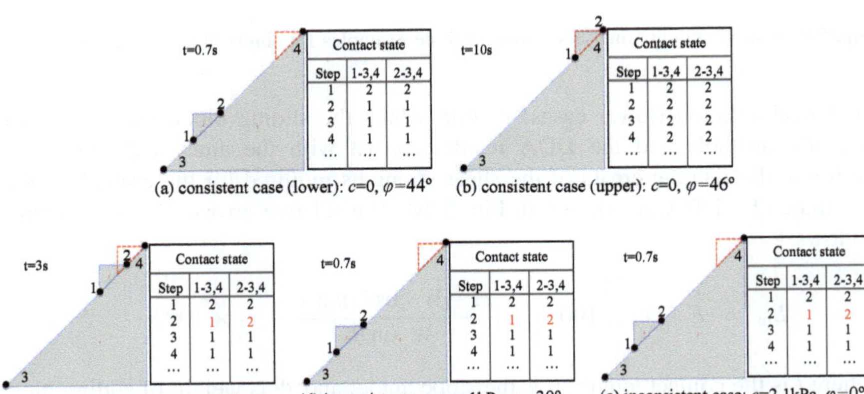

Fig. 5.11 Movement results and contact states of the original DDA for several cases

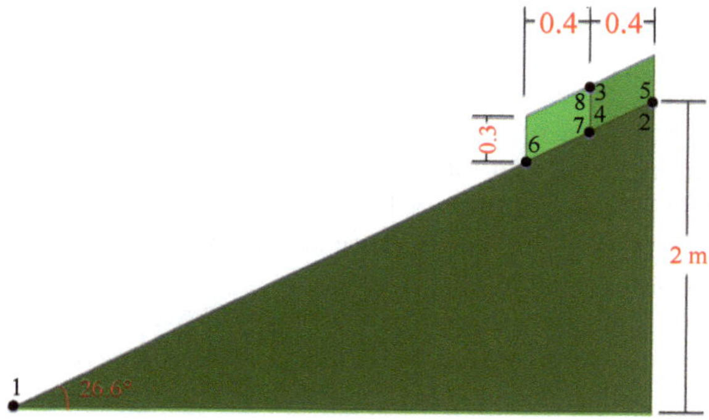

Fig. 5.12 Example of two blocks on incline

observed in all of examined inconsistent cases and not observed in all of consistent cases, no exception.

Hence, the treatment of edge-to-edge contact is a limitation of the original DDA, and it may be the reason of existence of so many inconsistent cases.

(b) *Two contiguous blocks on incline*

In order to further show the limitation of the original DDA, another example, two blocks on incline, is examined in this subsection. In this example, two alongside blocks are resting on a fixed block with a 26.57° incline (1V:2H) (Fig. 5.12). The schematic model and geometrical parameter are shown in Fig. 5.12. Also, the number of vertexes is shown in schematic model. The physical parameters of the blocks and the control parameters are the same with the examples of the single block on incline shown in previous subsection (Table 5.1).

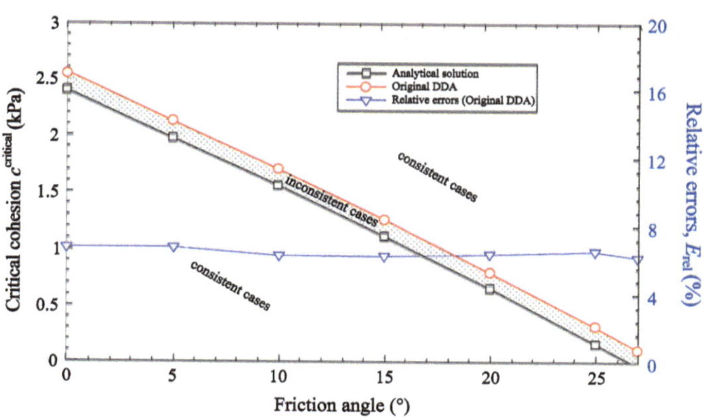

Fig. 5.13 Shear strengths required for the critical stability calculated by DDA and relative error for the example of two blocks on incline

Figure 5.13 shows the comparison of the critical stability analysis results between the original DDA and the analytical method, in which different cohesion is obtained for different given friction angles. The same with the examples of single block on incline, the DDA results are shown to be larger than the analytical results. In addition, the relative error of the shear strengths required for the critical stability calculated by DDA is also shown in Fig. 5.13. It can be found the relative error range between 6 and 7%. The relative error seems to be acceptable, but some other unreasonable phenomena should be noted.

For most cases (consistent zone in Fig. 5.13), DDA can calculate the failure of the model accurately. Some unreasonable phenomena, however, were observed in some inconsistent cases (shaded area in Fig. 5.13). These phenomena are shown in Fig. 5.14. First, the two contiguous blocks are separated from each other as shown in Fig. 5.14a, although they should have the same movement state (move together or both stable), i.e., always contact with each other, as the shear strengths of the joints are the same. Second, unreasonable displacements are observed for the case shown in Fig. 5.14b. Comparing it with result of the consistent case shown in Fig. 5.14c, we can find that inter-friction angle of 26.6° induces larger displacement than that is induced by inter-friction angle of 26.5°.

In order to investigate whether the treatment of edge-to-edge contact in the original DDA has relationship with these unreasonable phenomena, contact states of edge-to-edge contact for those inconsistent cases are recorded. Tables 5.2 and 5.3 show the recorded contact states for two inconsistent cases shown in Fig. 5.14a, b, respectively. The contact state illustrates the stability and movement situation of the blocks. For the first inconsistent case ($c = 0$, $\varphi = 27°$), contact situation of edge-to-edge shown in Fig. 5.5 appears in 2–8 steps (shaded area in Table 5.2) and impacts the following contact states. Contact number illustrates whether the contiguous blocks contact with each other, 6 means contact, whereas 4 means separate. In Table 5.2, they separate from each other from the fourteenth step. For the second inconsistent case ($c = 0$, $\varphi = 26.6°$), different contacts for one edge-to-edge are also observed in the second step (shaded area in Table 5.3). On the other hand, contact situation shown in Fig. 5.5 is not found in any consistent case.

The results of two blocks on incline further explain that the treatment of edge-to-edge contact in the original DDA is reason of inaccurate results for those inconsistent cases. An extension should be done to solve this problem.

5.3.2 *Extension of the Original DDA by Importing Edge-to-Edge Contact*

In original DDA, two vertex-to-edge contacts of one edge-to-edge contact may have the different states at the same time. And the cohesion between the contact will be immediately removed once the state change from *locked* to *sliding*. Then, the

Fig. 5.14 Results of the original DDA

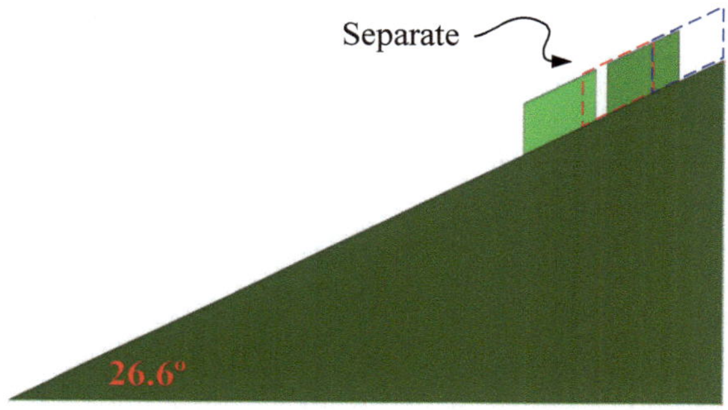

(a) inconsistent case: c=0kPa, φ=27°

(b) inconsistent case: c=0kPa, φ=26.6°

(c) consistent case: c=0kPa, φ=26.5°

improper *sliding* contact accelerates the failure of the other contact. After validation, this process may induce some unreasonable results. In order to solve this problem, re-evaluation of contact states of the whole edge-to-edge is added into the original DDA code. The purpose of the extension is to avoid the different contact states for one edge-to-edge contact, i.e., the situation shown in Fig. 5.3. In the extension, once two vertex-to-edge contacts of one joint have the different contact

Table 5.2 Contact state of an inconsistent case (c = 0 kPa, φ = 27°) of the original DDA

Step	Contact number	Original DDA					
		4,5-to-1,2		6,7-to-1,2		3,4-to-7,8	
		4-1,2	5-1,2	6-1,2	7-1,2	3-7,8	4-7,8
1	6	2	2	2	2	2	2
2-4	6	2	1	2	1	0	0
5-6	6	1	2	1	2	0	0
7	6	2	2	2	1	1	0
8	6	1	1	1	2	0	0
9	6	1	1	1	1	0	0
10-11	6	2	2	1	1	0	0
12	6	1	1	1	1	0	0
13	6	2	2	1	1	0	0
14	4	2	2	1	1	×	×
15	4	1	1	1	1	×	×
16-17	4	2	2	1	1	×	×
18-24	4	1	1	1	1	×	×
25-111	6	1	1	1	1	0	0
112	4	1	1	1	1	×	×
…	…	…	…	…	…	…	…

Note Contact states, 0, 1, and 2 mean open, sliding, and locked, respectively. Symbol × means not contact

states, an additional evaluation is added to recalculate the stability situation of the whole joint base on Mohr–Coulomb failure criterion as the follow criteria:

$$
\begin{aligned}
&\text{if } CS_{(P1\text{-}P3P4)} = CS_{(P2\text{-}P3P4)}\\
&\quad CS_{(P1P2\text{-to-}P3P4)} = C_{S(P1\text{-}P3P4)}.\\
&\text{if } CS_{(P1\text{-}P3P4)} \neq CS_{(P2\text{-}P3P4)}\\
&\quad \text{if } k_n[d_{n(P1\text{-}P3P4)} + d_{n(P2\text{-}P3P4)}]/2 > 0\\
&\qquad \text{and } k_s[d_{s(P1\text{-}P3P4)} + d_{s(P2\text{-}P3P4)}]/2 \le k_n[d_{n(P1\text{-}P3P4)} + d_{n(P2\text{-}P3P4)}]/2 \tan\varphi + cl_{P1P2}\\
&\qquad CS_{(P1P2\text{-to-}P3P4)} = \mathit{Locked}.\\
&\quad \text{if } k_n[d_{n(P1\text{-}P3P4)} + d_{n(P2\text{-}P3P4)}]/2 > 0\\
&\qquad \text{and } k_s[d_{s(P1\text{-}P3P4)} + d_{s(P2\text{-}P3P4)}]/2 > k_n[d_{n(P1\text{-}P3P4)} + d_{n(P2\text{-}P3P4)}]/2 \tan\varphi + cl_{P1P2}\\
&\qquad CS_{(P1P2\text{-to-}P3P4)} = \mathit{Sliding}.\\
&\quad \text{if } k_n[d_{n(P1\text{-}P3P4)} + d_{n(P2\text{-}P3P4)}]/2 \le 0\\
&\qquad CS_{(P1P2\text{-to-}P3P4)} = \mathit{Open}.
\end{aligned}
$$

where CS is abbreviation of contact state, subscripts of P_1-P_3P_4 and P_2-P_3P_4 are contacts of vertex-to-edge P_1-to-P_3P_4 and P_2-to-P_3P_4, respectively, and P_1P_2-to-P_3P_4 is contact of edge-to-edge P_1P_2-to-P_3P_4.

Finally, one joint between two blocks only has one contact state: (a) *open*, (b) *sliding*, or (c) *locked* (Fig. 5.15). It is different from the original DDA.

Table 5.3 Contact state of an inconsistent case ($c = 0$ kPa, $\varphi = 26.6°$) of the original DDA

Step	Contact number	Original DDA					
		4,5-to-1,2		6,7-to-1,2		3,4-to-7,8	
		4-1,2	5-1,2	6-1,2	7-1,2	3-7,8	4-7,8
1	6	2	2	2	2	2	2
2	6	2	1	2	1	0	0
3	6	1	1	1	1	0	0
…	…	…	…	…	…	…	…

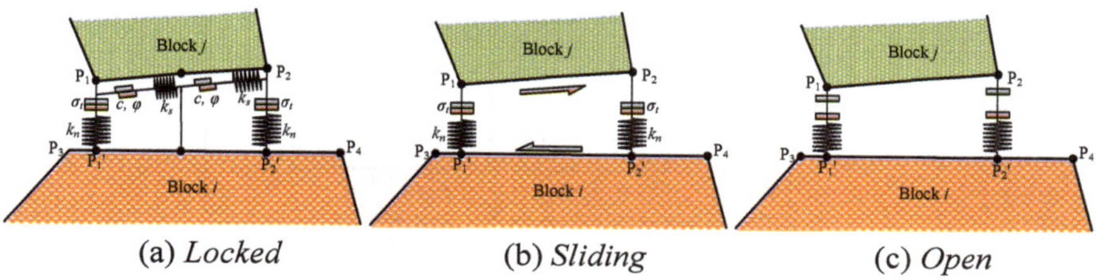

Fig. 5.15 Three states of edge-to-edge contact

A flowchart outlining the procedures of the present DDA program is shown in Fig. 5.16. The content in red box is the difference between the present DDA and the original DDA.

5.3.3 Validation of Static Sliding Block

(a) *Single block on incline*

The first kind of validation example is a single block on incline. Three examples shown in Fig. 5.7 are considered. Combinations of cohesion c and inter-friction angle φ that are required for the critical stability of the resting block were calculated by the present DDA. The failure criterion and the parameters remain unchanged with those in Table 5.1.

Comparisons of the results of critical shear strength from the original DDA, the present DDA, and the analytical solution are shown in Fig. 5.17, which shows that the results from the present DDA are almost identical to the analytical solution as the relative errors are less than 1% (Eq. 5.14).

In addition, the movement results and contact states of four cases from the original DDA and the present DDA are shown in Fig. 5.18. For first two cases (Fig. 5.18a–d), theoretical critical shear strength, (i) $c = 0$ and $\varphi = 45°$, and (ii) $c = 2$ kPa and $\varphi = 0$, was set. In Fig. 5.18a, b, the movement results of the

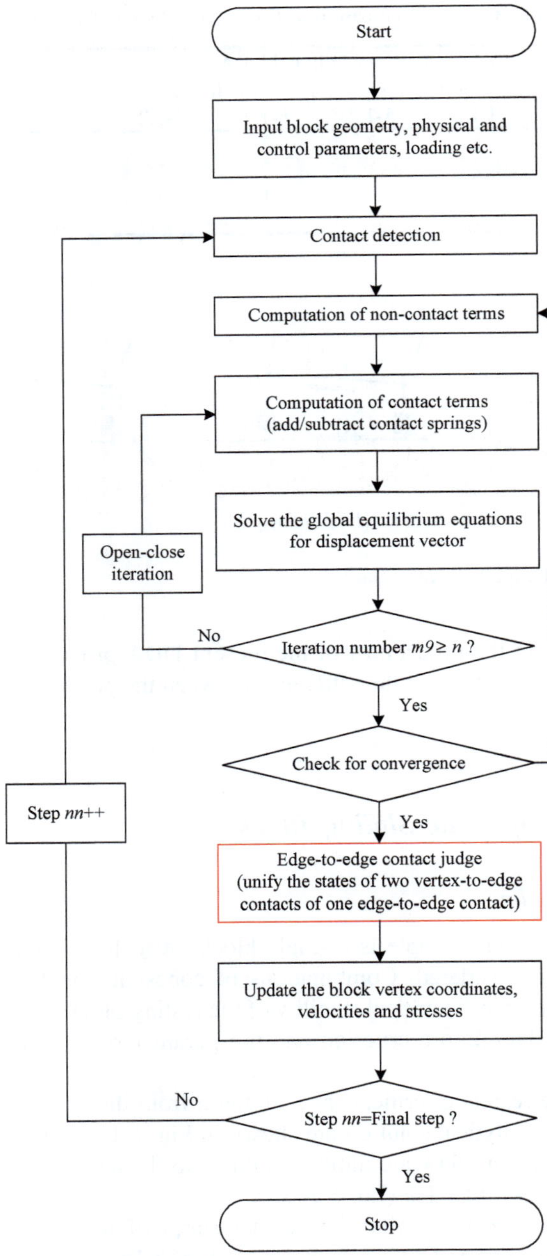

Note: n is defined by the user, default as 6

Fig. 5.16 Flowchart of the procedures of the present DDA program

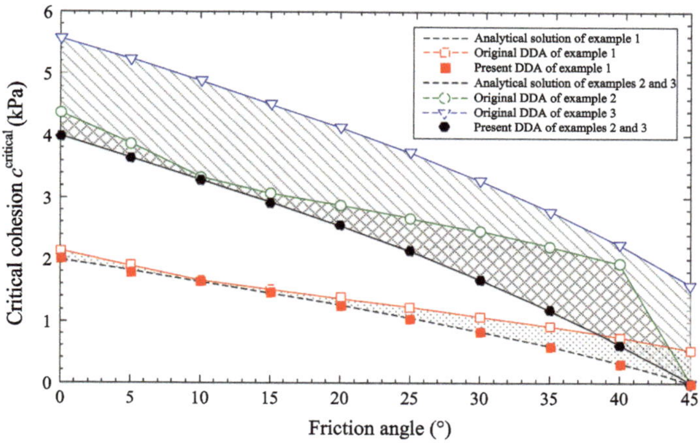

Fig. 5.17 Comparisons of the results of critical shear strength from the original DDA, the present DDA, and the analytical method

original DDA show that the resting block is unstable at the theoretical critical state, and contact states of joint in the original DDA are unreasonable (the red number in Fig. 5.18a, b). For the same cases, Fig. 5.18c, d shows the results of the present DDA, in which the resting block is stable and contact states are reasonable. In order to further investigate the accuracy of the present DDA, another two cases with the shear strength of a little less than the theoretical value ($c = 0$, $\varphi = 44.9°$ and $c = 1.99$ kPa, $\varphi = 0$) are validated. Figure 5.18e, f shows the results of the present DDA, in which stability situations, run-out situations, and contact states are consist well with the theoretical results.

(b) *Two contiguous blocks on incline*

Another validation example is two contiguous blocks on incline. The schematic model, physical parameters, and control parameters are the same with the example shown in Fig. 5.12. As shown in Sect. 5.3, 6–7% relative error of the critical shear strength and some unreasonable phenomena can be observed in the results of the original DDA. Same calculation was carried out by the present DDA.

Comparisons of the results of critical shear strength of the original DDA, the results of the present DDA, and the analytical solution are shown in Fig. 5.19, which shows that the results from the present DDA are almost identical to the analytical solution. And the relative errors are also shown in Fig. 5.19. It can be found that the relative error is changed from 6 to 7% of the original DDA to <1% of the present DDA.

In addition, the movement results of three cases from the original DDA and the present DDA are shown in Fig. 5.20. Some unreasonable phenomena of the results of the original DDA were described in last section (Fig. 5.20a, c, e). For the same cases, the results of present DDA are shown in Fig. 5.20b, d, f, respectively. In Fig. 5.20, we can find that the present DDA can avoid those unreasonable

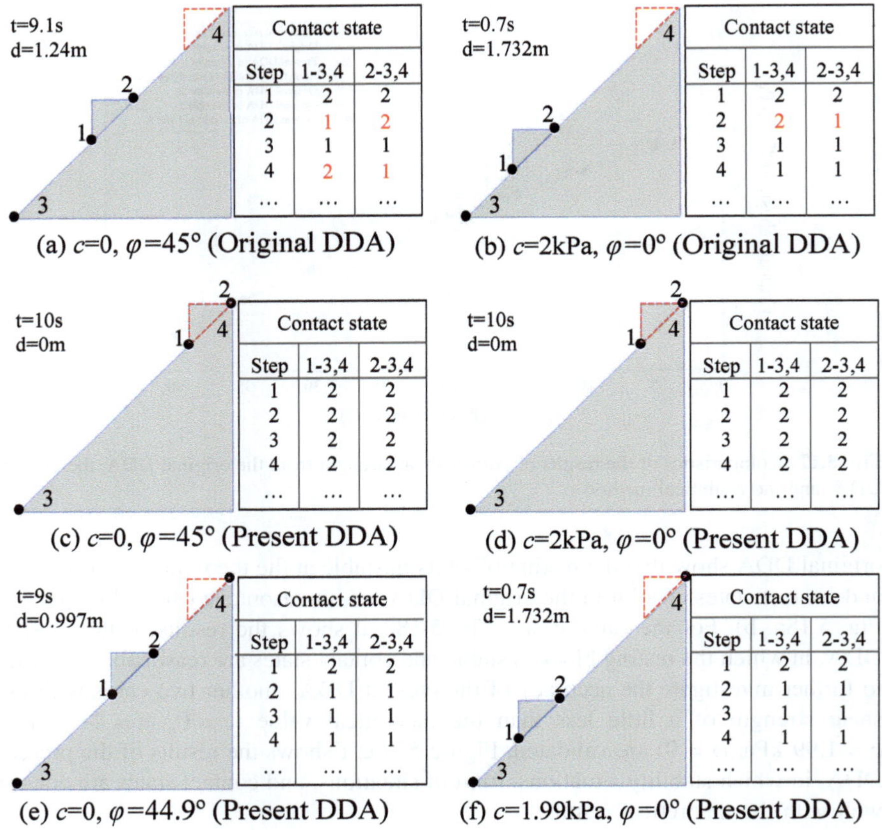

Fig. 5.18 Comparison of the calculated results of the original DDA and the present DDA

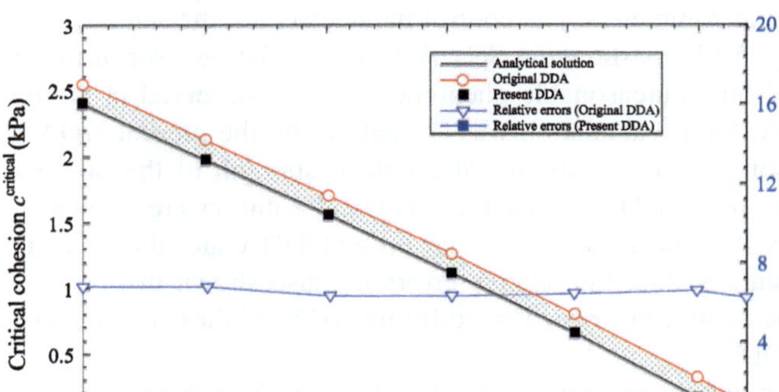

Fig. 5.19 Comparisons of the results of critical shear strength and relative error from the original DDA, the present DDA, and the analytical method

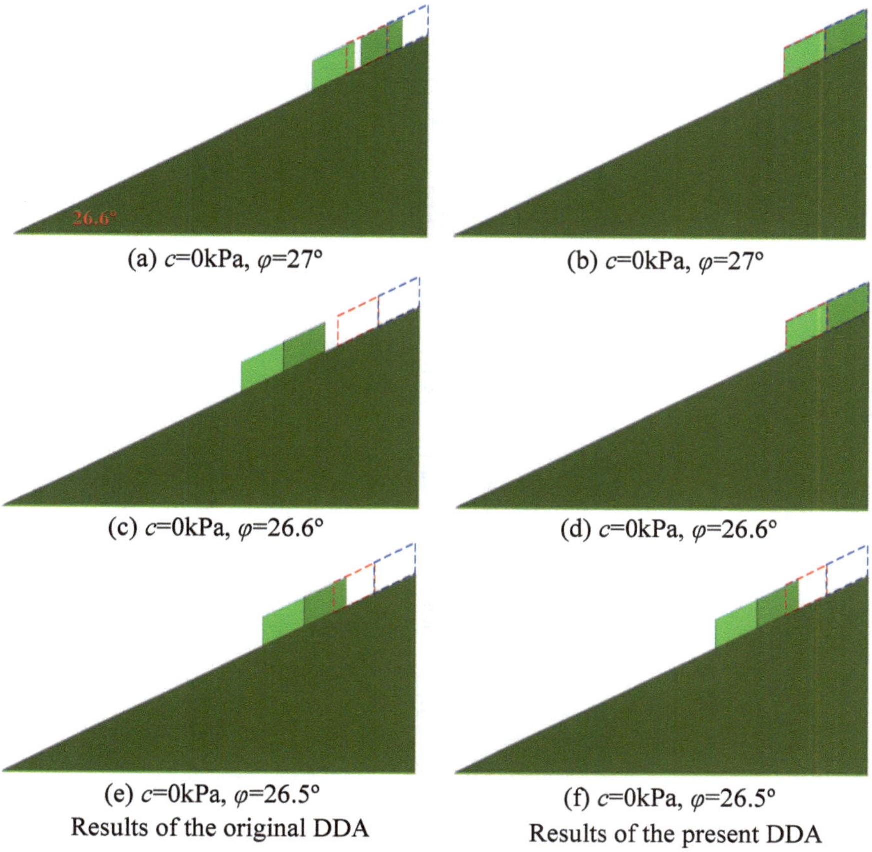

Fig. 5.20 Comparison of the results of the original DDA and the present DDA

phenomena and provide accurate results. And for two inconsistent cases, the contact states of the original DDA are given in Table 5.4, which shows that the present DDA can calculate the contact state accurately.

5.3.4 Remarks

Extension of the original DDA by importing an additional evaluation of edge-to-edge contact was presented in this paper, so it can be used to accurately investigate the failure behavior of joints dominated by both cohesion and inter-friction angle. The performance of the original DDA in deal with the combined cohesion-friction shear strength was examined through a series of classic examples, single block, and two contiguous blocks on incline. Conservative and

Table 5.4 Contact state of two inconsistent cases of the present DDA

Step	$c = 0$ kPa, $\varphi = 27°$							$c = 0$ kPa, $\varphi = 26.6°$						
	Contact number	4,5-to-1,2		6,7-to-1,2		3,4-to-7,8		Contact number	4,5-to-1,2		6,7-to-1,2		3,4-to-7,8	
		4-1,2	5-1,2	6-1,2	7-1,2	3-7,8	4-7,8		4-1,2	5-1,2	6-1,2	7-1,2	3-7,8	4-7,8
1	6	2	2	2	2	2	2	6	2	2	2	2	2	2
2	6	2	2	2	2	0	0	6	2	2	2	2	0	0
…	…	…	…	…	…	…	…	…	…	…	…	…	…	…

even unreasonable results were observed from the original DDA, especially when cohesion is considered. In the original DDA, every joint (edge-to-edge contact) is treated as two vertex-to-edge contacts that may have different contact states along with different treatments of cohesion. After detailed investigations and validations, this treatment of joint was found to be the key reason of the original DDA cannot simulate the cohesive material accurately. Re-evaluation of the contact state of the whole edge-to-edge was added into the original DDA to avoid different states of joint. In the extension, once the two vertex-to-edge contacts of one joint have different contact states, recalculation of the stability for the joint will be done based on Mohr–Coulomb failure criterion. At last, one joint only has one contact state. The novel feature of the present DDA is that it can automatically calculate and record the states of all edge-to-edge contacts, and thus, those edge-to-edge contacts with two states will be re-evaluated.

Some examples were used to validate the accuracy of the present DDA. The results of validation examples show that the present DDA can accurately simulate the failure process of cohesive-frictional material. The calculated shear strengths required for critical stability were almost identical to the analytical solution (relative error less than 1%). And those unreasonable phenomena appeared in the original DDA results were not observed in the present DDA results.

5.4 Validation of Dynamic Block Model

Some examples of six dynamic sliding block models (Chap. 4) were calculated by the rigorous dynamic block method. The same models are calculated using the DDA by the same excitations in this validation. The Young's modulus and Poisson's ratio of the blocks are chosen as $E = 1$ GPa and $\mu = 0.25$, respectively, which will be used in all the examples in the validation (Zhang et al. 2013). Table 5.5 shows other parameters in DDA program, and they are selected as reference the existing parameter used by Hatzor et al. (2004) and Ning and Zhao (2012). These parameters will be used throughout this paper. Extensive discussions on the influence of these parameters on the dynamic sliding modeling results in the DDA can be found in the papers by Hatzor and Feintuch (2001) and Tsesarsky et al. (2005).

Table 5.5 Parameters in DDA

Parameter	Value
Static or dynamic parameter, gg	1
Normal contact spring stiffness, $g0$	$10E$
Assumed maximum displacement ratio, $g2$	0.001
Step time, $g1$	0.0025 s

Six loading methods are carried out to investigate their difference in dynamic sliding prediction by both BLOCK and DDA. The relative difference (RD) between a DDA result and a theoretical result (BLOCK solution) is defined as follows:

$$\mathrm{RD} = \left|\frac{\mathrm{result_{DDA}} - \mathrm{result_{BLOCK}}}{\mathrm{result_{BLOCK}}}\right| \times 100\% \tag{5.15}$$

Table 5.6 shows the calculated permanent displacements and relative difference RD between DDA and BLOCK. In order to compare the difference between two corresponding cases a and b, the varying percentage $\Delta_{\mathrm{a-b}}$ of the sliding displacement between them is defined as follows:

$$\Delta_{\mathrm{a-b}} = \frac{d_{\mathrm{b}} - d_{\mathrm{a}}}{d_{\mathrm{a}}} \times 100\% \tag{5.16}$$

where d_{a} and d_{b} are calculated residual displacements for cases a and b. This varying percentage $\Delta_{\mathrm{a-b}}$ is also shown in Table 5.6.

From the RD in Table 5.6, we can find that the DDA results match the BLOCK theoretical results well. For all calculated cases, most of the RD are smaller than 5%, and the maximum relative displacement $\mathrm{RD_{max}} < 8.5\%$.

Figure 5.21 shows the velocity and displacement comparisons of the sliding block between the DDA and the theoretical solutions. Generally, the DDA results match the theoretical results well. From the acceleration results, we can found that there are the ups and downs in the DDA result. The reason for this phenomenon is that the penalty method is used in the DDA. But its influences on the block velocity and displacement are insignificant because the small time step or high frequency of acceleration itself.

5.4.1 *Effects of Model Types on the Residual Displacement*

When the seismic loading parallelly acts on the sliding block (model 1) or base block (model 2) in the same direction, it induced permanent displacements are very different among different excitations. This point can be found in Table 5.6, in which the varying percentage Δ_{1-2} for all cases is changed from −25.24 to 98.16%. Similar results also can be found from other model group, Δ_{3-4} of models 3 and 4 and Δ_{5-6} of models 5 and 6. In addition, when varying the acting block along with the direction of seismic loading [e.g., the cases of Pacoima Dam—254° (opposite) in Table 5.6], the calculated displacements are equivalent.

Figure 5.22 shows the comparison of velocity and displacement time histories from one case of model 1 (middle case in Fig. 5.22) and two cases of model 2 (two directions of seismic loading apply).

Table 5.6 Permanent displacements of six models under different excitations

	Ricker wavelet			Sinusoidal			Pacoima Dam - 164°			Pacoima Dam - 254°			Pacoima Dam - 254° (opposite)		
	BLOCK	DDA	RD(%)	BLOCK	DDA	RD(%)	BLOCK	DDA	RD(%)	BLOCK	DDA	RD(%)	BLOCK	DDA	RD(%)
Model 1	0.361	0.331	8.343	0.913	0.896	1.818	0.369	0.364	1.490	0.197	0.191	3.146	0.264	-	-
Model 2	0.417	0.397	4.887	1.809	1.766	2.377	0.335	0.313	6.679	0.264	0.254	3.719	0.197	-	-
Model 3	0.488	0.498	2.173	1.117	1.100	1.522	0.558	0.529	5.213	0.280	0.278	0.714	0.358	0.373	4.157
Model 4	0.514	0.498	3.016	2.475	2.281	7.838	0.497	0.462	7.041	0.358	0.354	1.228	0.280	0.283	1.106
Model 5	0.581	0.562	3.338	1.265	1.246	1.502	0.584	0.592	1.473	0.317	0.322	1.705	0.314	0.339	8.193
Model 6	0.586	0.563	4.008	3.008	2.798	6.981	0.538	0.512	4.886	0.427	0.407	4.636	0.251	0.251	0
$\Delta_{1\text{-}2}$	15.69	20.05		98.16	97.03		-9.15	-13.94		33.69	32.90		-25.24	-	
$\Delta_{3\text{-}4}$	5.37	0.02		121.58	107.36		-10.95	-12.66		27.84	27.17		-21.82	-24.11	
$\Delta_{5\text{-}6}$	0.91	0.21		137.79	124.56		-7.78	-13.56		34.82	26.41		-20.05	-15.56	
$\Delta_{3\text{-}5}$	19.15	12.72		13.25	13.27		4.57	11.94		13.06	15.82		-12.47	-9.08	
$\Delta_{4\text{-}6}$	14.11	12.94		21.54	22.67		8.29	10.80		19.24	15.12		-10.49	1.16	
$Times_{max}$	1.63	1.70		3.29	3.12		1.74	1.89		2.17	2.13		1.82	-	

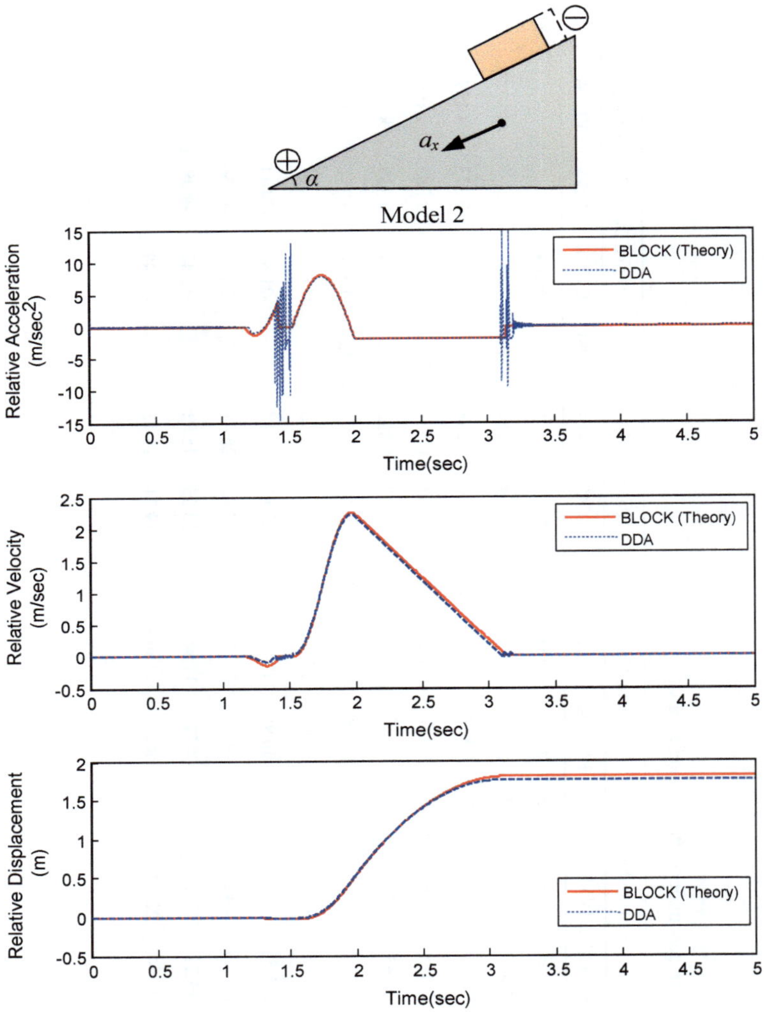

Fig. 5.21 Kinematic parameters, *a*, *v*, and *d* comparisons between the DDA and the BLOCK theoretical solutions

Differences among six different sliding block models are significant. Generally, the results of permanent displacement from model 1 are often the smallest one among all models in same geometric and loading situation (blue word in Table 5.6). And the largest displacements are occurred in the model 6 (red word in Table 5.6). The times between the largest and the smallest displacement are as large as 3.29 for sinusoidal excitation.

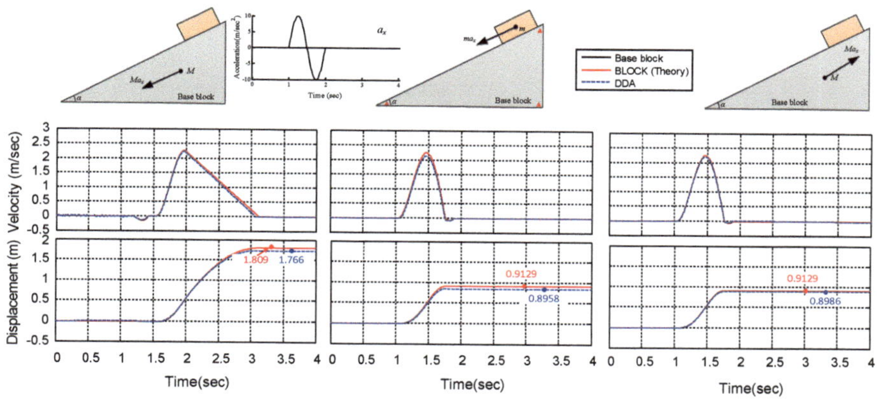

Fig. 5.22 Comparison of velocity and displacement time histories from three cases

As an example, Fig. 5.22 gives the block relative sliding velocity and displacement time histories to the base corresponding BLOCK theoretical results for the models 1 and 6 under the real earthquake loading from Pacoima Dam—254° records in San Fernando earthquake. Both velocity and displacement histories of DDA are consist well with those of BLOCK theoretical solution. The RDs for residual displacements of models 1 and 6 are 3.146 and 4.636%, respectively. In addition, for BLOCK theoretical solution, the maximum displacement from model 6, 0.427 m, is 2.17 times of that from model 1, 0.407 m, and 2.13 times for DDA solution in the same models.

5.4.2 Effects of Vertical Seismic Force on the Residual Displacement

In this section, comparisons between models with or without vertical seismic component are carried out to investigate its effect on dynamic sliding displacement. The varying percentage Δ_{a-b} of the sliding displacement for cases a and b is used to evaluate the influence of vertical seismic force; in here, subscript b represents the case of a vertical component is applied combined with the horizontal seismic loading and subscript a represents the case of only horizontal seismic loading is applied (Fig. 5.23).

From Δ_{a-b} in Table 5.6, we can find that the excitation acting on either the sliding block or the base block, and the effect of vertical seismic force on residual displacement may be significant under various excitations. In calculated cases, the maximum difference occurred in models 4 and 6 under excitation of sinus plus, and it reaches as large as 21.54 and 22.67% from BLOCK and DDA, respectively. This

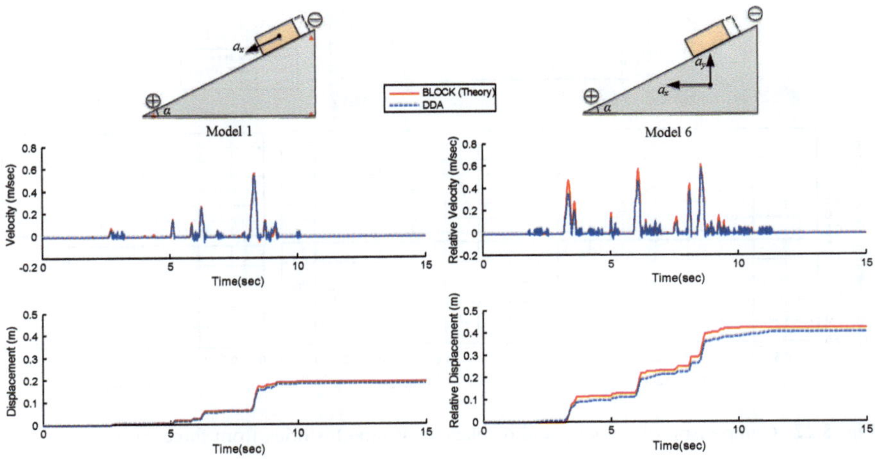

Fig. 5.23 Different permanent displacements calculated from different models

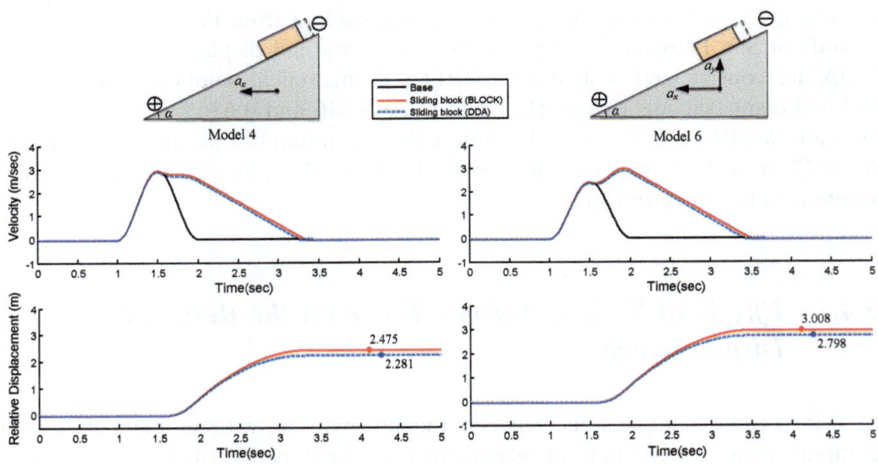

Fig. 5.24 Effects of vertical seismic loading on induced permanent displacement

maximum difference is shown in Fig. 5.24. It should be noted that the used idealized vertical seismic force only the half of the horizontal component, while peak ground acceleration of vertical component (PGA_V) recorded from an real earthquake event often reaches the horizontal peak ground acceleration (PGA_H), even exceeds the PGA_H. Hence, the effects of vertical seismic force on sliding system should be considered.

5.4.3 *Effects of Model Strike Direction on the Residual Displacement*

In the previous examples, zero degree of model strike direction, $\eta = 0$, was used to verify the DDA method for dynamic sliding modeling. In this subsection, the model located in different strike directions is examined by the DDA and the BLOCK program. As a preliminary study, two directions are considered, $\eta = 0$ and $\eta = 180$, i.e., two models have the opposite direction. A symmetrical block sliding model is constructed as shown in Fig. 5.25. The real seismic acceleration record of Pacoima Dam, 254°, is horizontally–vertically loaded to the base block (model 6). The relative velocity time histories of the two blocks to the base along the slope surface are plotted in Fig. 5.25. The DDA results match the BLOCK ones well. The two blocks located on the two slides of the base have obviously different dynamic behaviors. Under the earthquake loading, two blocks slide and stop on the base repeatedly and finally rest after the earthquake loading. Relative displacement time histories of the two blocks to the base along the slope surface are also shown in Fig. 5.25. The DDA results agree quite well with the BLOCK results. The RDs between the DDA results and the BLOCK results of the two blocks are 0.000 and 4.636%, respectively.

Block 2 on the right slide has larger relative displacement than block 1 on the left surface. After the blocks stop finally, the values of relative displacement of the two blocks are about 0.25 and 0.40 m.

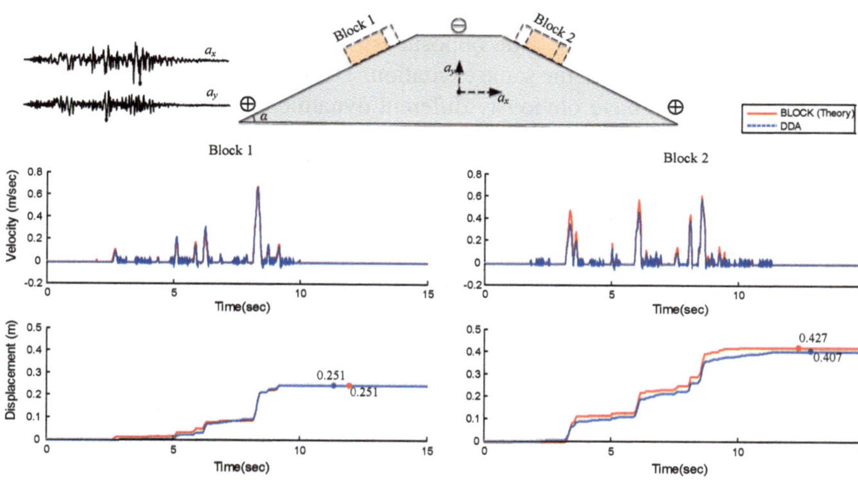

Fig. 5.25 Effects of slope direction on induced permanent displacement

5.4.4 Remarks

In this section, a detailed investigation of near-fault earthquake loading induced displacement of sliding system is carried out by extended discontinuous deformation analysis (DDA). Six different models under excitations of two idealized wavelets and one group of accelerograms are considered. A unified rigorous analytical program called BLOCK for all models is proposed in which ground shakings can be applied on base block or sliding block in direction of either horizontal or parallel to the inclined plane, with or without a simultaneous vertical component of motion. Both analytical solution and DDA simulation results for 6 models are presented. Comparisons of the theory solution and DDA results for different cases are carried out. From the findings of this study, the following conclusions can be drawn:

(1) The results show that DDA can simulate the movement of dynamic block successfully and give accurate results. For all calculated cases, RDs between the DDA results and the BLOCK results are small. The DDA results match the theoretical results well.
(2) Model type of sliding system has a significant influence on the dynamic behavior of sliding block. If the vertical seismic force is not considered, acting the horizontal seismic excitation to the base is equivalent to acting the seismic excitation to the sliding block in the opposite directions, whereas this equation is invalid once the vertical seismic force is considered, only if the vertical seismic force also changes the direction.
(3) The vertical seismic force should be considered because it may have significant effect on the block sliding displacements.
(4) Two symmetrical blocks with opposite strike direction have asymmetrical dynamic behavior under the same excitation. The two blocks located on the two slides of the base have obviously different dynamic behavior.

5.5 Validation of Trampoline Effects

A four-layer model shown in Fig. 5.26 is used for clarifying the mechanism of generating extreme waveform.

The parameters are listed in Table 5.7. Since the bottom block represents the bed rock, a large density is taken for large mass.

Two types of vertical seismic wave loadings act in the center point of the bottom block. The one type is regular sine waveform with amplitude of 6–8 m/s^2 and frequence of 0.5–10 Hz. The other is the real waveform recorded by the downhole sensor of the West Ichinoseki station.

The response waveform of the top block estimated by DDA simulation is shown in Fig. 5.27 for the regular sine waveform with amplitude of 8 m/s^2 and frequency of 8 Hz. It can be seen that the peak response reaches as larger as 5 g.

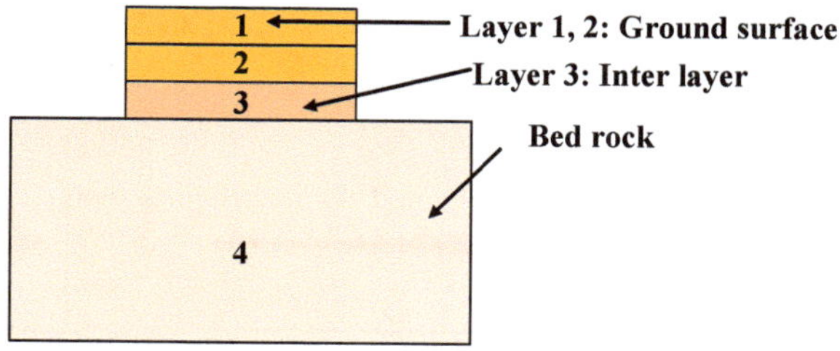

Fig. 5.26 A four-layer model used in DDA simulation

Table 5.7 Physical parameters and control parameters of four-layer model used in DDA simulation

Parameter	Value		
	Layer 1, 2	Layer 3	Layer 4
Density (Kg/m^3)	2000	2000	2000 × 10^3
Young's modulus (Pa)	10^9	10^7	10^9
Poisson' ratio	0.1	0.2	0.3
Penalty	10^8		
Time step (s)	0.001		

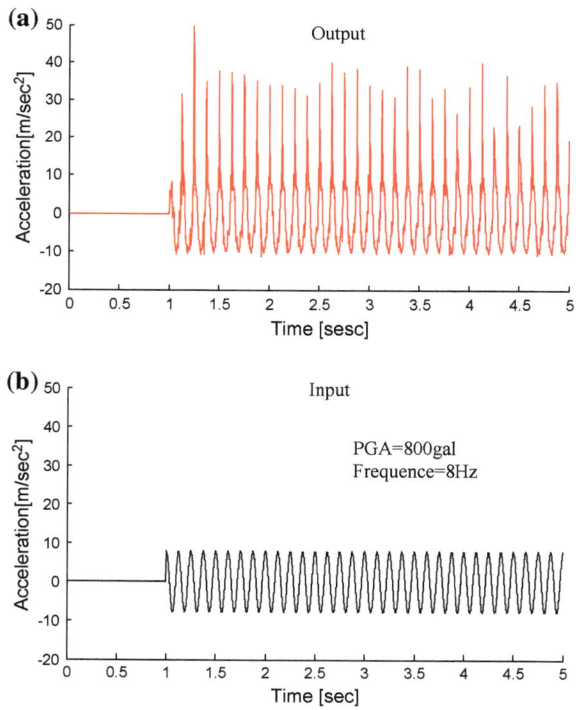

Fig. 5.27 Response waveform of the top block simulated by DDA (regular waveform)

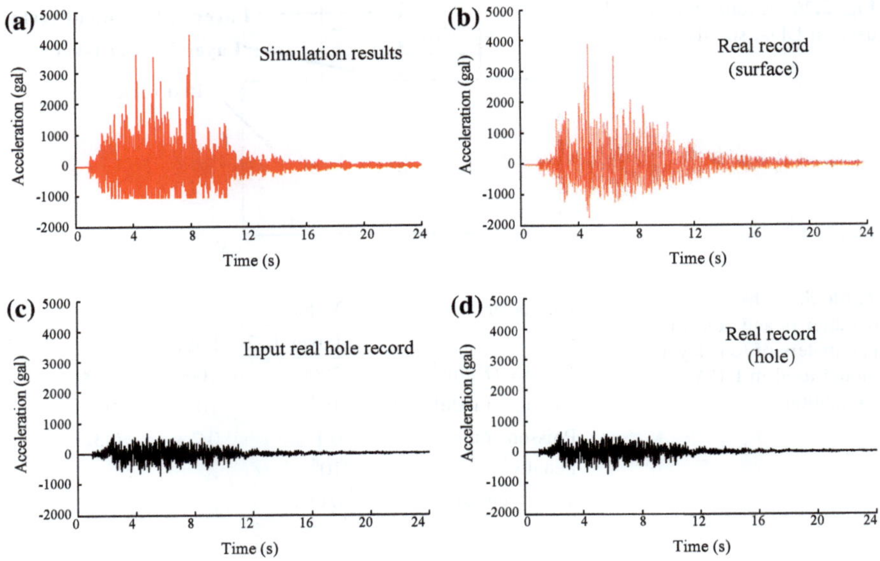

Fig. 5.28 Response waveform of the top block simulated by DDA (real record)

The result for the downhole real waveform input is shown in Fig. 5.28, comparing the response waveform obtained from DDA to the real waveform recorded by the surface sensor. It can be seen that there are some similarities and differences. The peak acceleration values are the same order, but the distributions are little different. This is because the real ground is different from the simple model.

5.6 Simulation of the Donghekou Landslide by Using the Developed DDA Program

Donghekou rockslide is a typical large-scale rapid and long run-out rockslide triggered by the 2008 Wenchuan earthquake. Figure 4.2 shows (a) an air photograph and (b) I-I section of the Donghekou rockslide. Donghekou village, where about 1500 people were lived, was located at the convergence of the Hongshi River and Xiasi River. The landslide destroyed all of the houses in the village and killed about 780 people. The rockslide has a height difference between the toe and main scarp of 700 m, a slide distance of 2400 m. The average thickness of the slide mass in the source area is about 80 m, with the maximum thickness of 110 m, and the total volume of the Donghekou rockslide is about 10–69 million m^3 (Yin et al. 2009; Wang et al. 2013). The slide mass of the Donghekou rock avalanche is mainly composed of Sinian limestone and dolomite limestone, together with Cambrian carbonaceous slate and phyllite (Sun et al. 2011). In the source area, there was a lot of sliding mass left covering the sliding surface of the rockslide, making it difficult to determine the material properties in the slide zone. Hence, how to determine the shear strength of the material is an important and primary question in the sliding analysis.

5.6.1 DDA Model, Parameters, and Seismic Loadings

Based on the profile section shown in Fig. 4.2b, a real size DDA model (Fig. 5.29) was built, in which two kinds of material were set, the sliding mass and the base block. Voronoi method was used to segment the sliding mass into small blocks. The base block was fixed by setting several fixed points (the red triangle in the model).

The parameters of discontinuity, cohesion c and inter-friction angle φ, are variable in a possible range. It should be noted that the inter-fiction angle is smaller than general value due to the scenario that could happen during the strong Wenchuan earthquake. The tensile strength σ_t is set as 0. Table 5.8 lists all of the parameters for the DDA simulation. The purpose of the study is to back-calculate the shear strength, c and φ, based on the stability situation and the run-out situation by the present DDA.

Figure 5.30 shows the input combined acceleration records from the MZQP strong motion station from the 2008 Wenchuan earthquake. Velocity and displacement time histories are obtained by once and twice integration from acceleration record, respectively. The horizontal-and-vertical acceleration records are acted to the base block at the same time.

5.6.2 Results of DDA Simulation

Figure 5.31 shows the post-failure behavior of the Donghekou landslide simulated by the seismic DDA code. Simulated results show that the sliding blocks run-out about 2400 m. After overlapping the final step of DDA calculation with the topographic cross section at the Donghekou landslide (Fig. 4.2b), the deposit pattern of the simulated Donghekou landslide under horizontal-and-vertical situation coincides well with local topography.

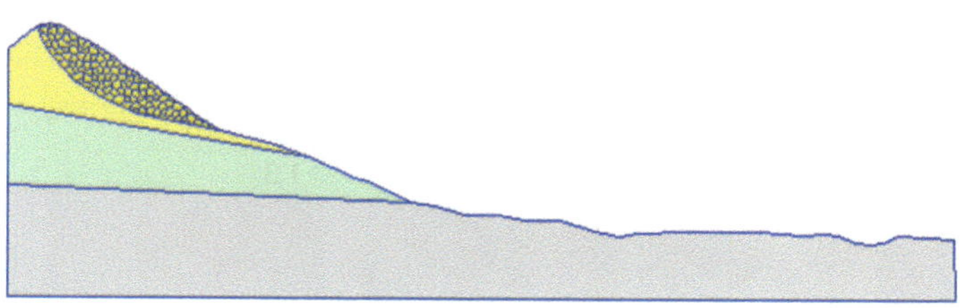

Fig. 5.29 DDA model of the Donghekou rockslide

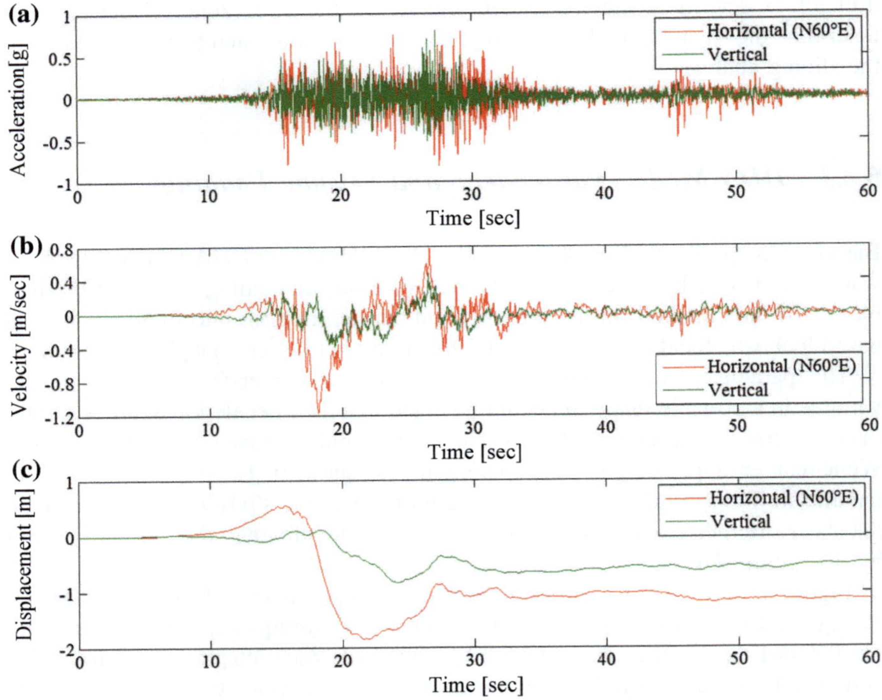

Fig. 5.30 Input horizontal and vertical ground acceleration records

Table 5.8 Parameters for the DDA simulation

Parameter		Value	
		Sliding mass	Base block
Physical	Unit weight, γ (kN/m^3)	21.5	25
	Elasticity modulus, E (GPa)	5.4	5.4
	Poisson's ratio, v	0.28	0.1
Discontinuity	Inter-friction angle, φ (°)	15	
	Cohesion, c (MPa)	2.1	
	Tensile strength, σ_t (kPa)	0	
Control	Dynamic control parameter	1	
	Maximum displacement ratio, g_2	0.001	
	Time step, g_1 (s)	0.01	
	Contact spring stiffness, k_n (kN/m)	1.0×10^6	
	Factor of over-relaxation	1.3	

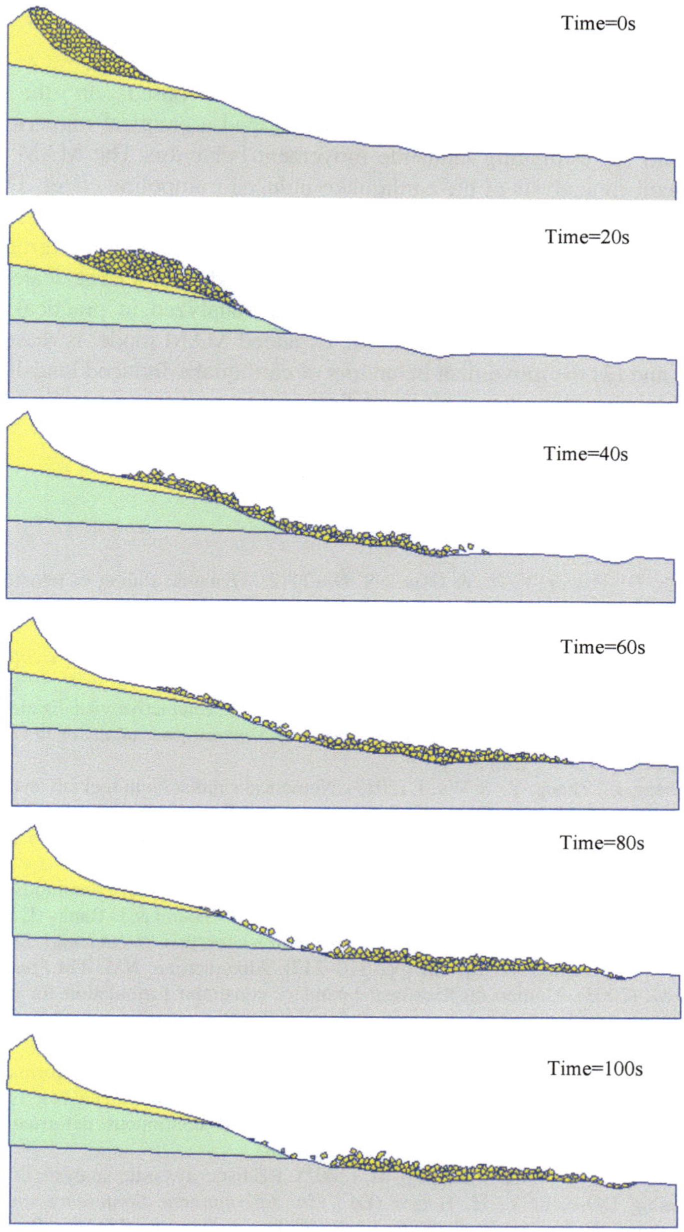

Fig. 5.31 DDA simulation results of the run-out of the Donghekou landslide

5.7 Conclusions

This chapter presented a long run-out model based on the so-called earthquake-induced trampoline effect and developed a practical numerical simulation program for estimating landslide movement behaviors The MAM is derived from mechanism analysis of the earthquake-induced trampoline effect. The original DDA was extended. A practical numerical simulation program was developed by incorporating the MAM into the extended DDA. After an extreme ground movement with the PGA of 4000 gal is successfully reproduced, a large-scale landslides induced by the 2008 Wenchuan earthquake are analyzed in practical numerical simulations. The results show that (1) the extended MAM model is reasonable and applicable and (2) the movement behaviors of earthquake-induced landslides can be analyzed by using the numerical simulation program.

References

Bakun-Mazor, D., Hatzor, Y. H., & Glaser, S. D. (2012). Dynamic sliding of tetrahedral wedge: The role of interface friction. *International Journal for Numerical and Analytical Methods in Geomechanics, 36*(3), 327–343.

Chen, G. (2003). Numerical modelling of rock fall using extended DDA. *Chinese Journal of Rock Mechanics and Engineering, 22*(6), 926–931.

Chen, G., Miki, S., & Ohnishi, Y. (1997). Development of the interactive visualization system for DDA. In *9th International Conference on Computer Methods and Advances in Geomechanics* (pp. 495–500). Wuhan, China.

Chen, G., Zheng, L., Zhang, Y., & Wu, J. (2013). Numerical simulation in rockfall analysis: A close comparison of 2-D and 3-D DDA. *Rock Mechanics and Rock Engineering, 46*(3), 527–541.

Cheng, Y. M. (1998). Advancements and improvement in discontinuous deformation analysis. *Computers and Geotechnics, 22*(2), 153–163.

Dong, X., Wu, A., & Ren, F. (1996). A preliminary application of discontinuous deformation analysis (DDA) to the Three Gorges Dam project. In M. R. Salami & D Banks (Eds.), *The First International Forum on Discontinuous Deformation Analysis (DDA) and Simulations of Discontinuous Media, Berkeley, CA* (pp. 310–317). Albuquerque, NM: TSI Press.

Doolin, D. M. (2005). Unified displacement boundary constraint formulation for discontinuous deformation analysis (DDA). *International Journal for Numerical and Analytical Methods in Geomechanics, 29*(12), 1199–1207.

Doolin, D. M., & Sitar, N. (2002). Displacement accuracy of discontinuous deformation analysis method applied to sliding block. *Journal of Engineering Mechanics, 128*(11), 1158–1168.

Doolin, D. M., & Sitar, N. (2004). Time integration in discontinuous deformation analysis. *Journal of Engineering Mechanics, 130*(3), 249–258.

Hatzor, Y. H., Arzi, A. A., & Tsesarsky, M. (2002). Realistic dynamic analysis of jointed rock slopes using DDA. In Y. H. Hatzor (Ed.), *5th International Conference on Analysis of Discontinuous Deformation—Stability of Rock Structures* (pp. 47–56). Rotterdam, The Netherlands: Abingdon, Balkema.

Hatzor, Y., Arzi, A. A., Zaslavsky, Y., & Shapira, A. (2004). Dynamic stability analysis of jointed rock slopes using the DDA method: King Herod's Palace, Masada, Israel. *International Journal of Rock Mechanics and Mining Sciences, 41*(5), 813–832.

Hatzor, Y. H., & Benary, R. (1997). The stability of a laminated voussoir beam: Back analysis of a historic roof collapse using DDA. *International Journal of Rock Mechanics and Mining Sciences, 35*(2), 165–181.

Hatzor, Y. H., & Feintuch, A. (2001). The validity of dynamic block displacement prediction using DDA. *International Journal of Rock Mechanics and Mining Sciences, 38,* 599–606.

Jiang, Q. H., & Yeung, M. R. (2004). A model of point-to-face contact for three-dimensional discontinuous deformation analysis. *Rock Mechanics and Rock Engineering, 37*(2), 95–116.

Kamai, R., & Hatzor, Y. H. (2008). Numerical analysis of block stone displacements in ancient masonry structures: A new method to estimate historic ground motions. *International Journal for Numerical and Analytical Methods in Geomechanics, 32*(11), 1321–1340.

Ke, T. C. (1996). The issues of rigid-body rotation in DDA. In *First International Forum on Discontinuous Deformation Analysis (DDA) and Simulations of Discontinuous Media* (pp. 318–325). Berkeley, USA.

Kim, Y.-I., Amadei, B., & Pan, E. (1999). Modeling the effect of water, excavation sequence and rock reinforcement with discontinuous deformation analysis. *International Journal of Rock Mechanics and Mining Sciences, 36*(7), 949–970.

Koo, C. Y., & Chern, J. C. (1998). Modification of the DDA method for rigid block problems. *International Journal of Rock Mechanics and Mining Sciences, 35,* 683–693.

Kottenstette, J. T. (1999). DDA analysis of the RCC modification for Pueblo Dam. In B. Amadei (Ed.), *3rd International Conference on Analysis of Discontinuous Deformation—From Theory to Practice (ICADD-3)* (pp. 127–132). Vail, CO: American Rock Mechanics Association, Alexandria.

Kveldsvik, V., Einstein, H. H., Nilsen, B., & Blikra, L. H. (2008). Numerical analysis of the 650,000 m^2 Åknes rock slope based on measured displacements and geotechnical data. *Rock Mechanics and Rock Engineering, 42*(5), 689–728.

Law, H. K., & Lam, I. P. (2003). Evaluation of seismic performance for tunnel retrofit project. *Journal of Geotechnical and Geoenvironmental Engineering, 129*(7), 575–589.

Lin, C. T., Amadei, B., Jung, J., & Dwyer, J. (1996). Extensions of discontinuous deformation analysis for jointed rock masses. *International Journal of Rock Mechanics and Mining Sciences & Geomechanics Abstracts, 33*(7), 671–694.

Ma, G. C., Matsuyama, H., Nishiyama, S., & Ohnishi, Y. (2011). Practical studies on rockfall simulation by DDA. *Journal of Rock Mechanics and Geotechnical Engineering, 3*(1), 57–63.

MacLaughlin, M. M., & Doolin, D. M. (2006). Review of validation of the discontinuous deformation analysis (DDA) method. *International Journal for Numerical and Analytical Methods in Geomechanics, 30*(4), 271–305.

MacLaughlin, M., Sitar, N., Doolin, D., & Abbot, T. (2001). Investigation of slope-stability kinematics using discontinuous deformation analysis. *International Journal of Rock Mechanics and Mining Sciences, 38,* 753–762.

Ning, Y., Yang, J., An, X., & Ma, G. (2010). Simulation of blast induced crater in jointed rock mass by discontinuous deformation analysis method. *Frontiers of Architecture and Civil Engineering in China, 4*(2), 223–232.

Ning, Y., Yang, J., An, X., & Ma, G. (2011a). Modelling rock fracturing and blast-induced rock mass failure via advanced discretisation within the discontinuous deformation analysis framework. *Computers and Geotechnics, 38*(1), 40–49.

Ning, Y., Yang, J., Ma, G., & Chen, P. (2011b). Modelling rock blasting considering explosion gas penetration using discontinuous deformation analysis. *Rock Mechanics and Rock Engineering, 44*(4), 483–490.

Ning, Y., & Zhao, Z. (2012). A detailed investigation of block dynamic sliding by the discontinuous deformation analysis. *International Journal for Numerical and Analytical Methods in Geomechanics*, 1–21.

Scheldt, T., Lu, M., & Myrvang, A. (2002). Numerical analysis of Gjovik cavern: A comparison of continuous and discontinuous results by using Phase 2 and DDA. In Y. H. Hatzor (Ed.), *5th International Conference on Analysis of Discontinuous Deformation—Stability of Rock Structures* (pp. 125–132). Rotterdam, The Netherlands: Abingdon, Balkema.

Shi, G.-H. (1988). *Discontinuous deformation analysis a new numerical model for the statics and dynamics of block systems*. Berkeley: University of California.

Shi, G.-H., & Goodman, R. E. (1985). Two dimensional discontinuous deformation analysis. *International Journal for Numerical and Analytical Methods in Geomechanics, 9,* 541–556.

Sitar, N., & Maclaughlin, M. M. (1997). Kinematics and discontinuous deformation analysis of landslide movement. Invited Keynote Lecture. In *II Pan-American Symposium on Landslides*. Rio de Janeiro.

Sitar, N., MacLaughlin, M. M., & Doolin, D. M. (2005). Influence of kinematics on landslide mobility and failure mode. *Journal of Geotechnical and Geoenvironmental Engineering, 131* (6), 716–728.

Sun, P., Zhang, Y., Shi, J., & Chen, L. (2011). Analysis on the dynamical process of Donghekou rockslide-debris flow triggered by 5.12 Wenchuan earthquake. *Journal of Mountain Science, 8* (2), 140–148.

Thomas, P. A., & Bray, J. D. (1999). Capturing nonspherical shape of granular media with disk cluster. *Journal of Geotechnical and Geoenvironmental Engineering, 125*(3), 169–178.

Tsesarsky, M., & Hatzor, Y. (2006). Tunnel roof deflection in blocky rock masses as a function of joint spacing and friction—A parametric study using discontinuous deformation analysis (DDA). *Tunnelling and Underground Space Technology, 21*(1), 29–45.

Tsesarsky, M., Hatzor, Y. H., & Sitar, N. (2005). Dynamic displacement of a block on an inclined plane: Analytical, experimental and DDA results. *Rock Mechanics and Rock Engineering, 38*(2), 153–167.

Wang, L.-Z., Jiang, H.-Y., Yang, Z.-X., Xu, Y.-C., & Zhu, X.-B. (2013). Development of discontinuous deformation analysis with displacement-dependent interface shear strength. *Computers and Geotechnics, 47,* 91–101.

Wu, J. (2004). Simulation of the mechanical behavior of inclined jointed rock masses during tunnel construction using Discontinuous Deformation Analysis (DDA). *International Journal of Rock Mechanics and Mining Sciences, 41*(5), 731–743.

Wu, J.-H. (2007). Applying discontinuous deformation analysis to assess the constrained area of the unstable Chiu-fen-erh-shan landslide slope. *International Journal for Numerical and Analytical Methods in Geomechanics, 31*(5), 649–666.

Wu, J.-H. (2010). Seismic landslide simulations in discontinuous deformation analysis. *Computers and Geotechnics, 37*(5), 594–601.

Wu, J.-H., & Chen, C.-H. (2011). Application of DDA to simulate characteristics of the Tsaoling landslide. *Computers and Geotechnics, 38*(5), 741–750.

Wu, J., Lin, J., & Chen, C. (2009). Dynamic discrete analysis of an earthquake-induced large-scale landslide. *International Journal of Rock Mechanics and Mining Sciences, 46*(2), 397–407.

Wu, J.-H., Ohnishi, Y., & Nishiyama, S. (2005). A development of the discontinuous deformation analysis for rock fall analysis. *International Journal for Numerical and Analytical Methods in Geomechanics, 29*(10), 971–988.

Yeung, M. (1993). Analysis of a mine roof using the DDA method. *International Journal of Rock Mechanics and Mining Sciences & Geomechanics Abstracts, 30*(7), 1411–1417.

Yeung, M. R., & Leong, L. L. (1997). Effects of joint attributes on tunnel stability. *International Journal of Rock Mechanics and Mining Sciences, 34*(3–4), 348.e1–348.e18.

Yin, Y., Wang, F., & Sun, P. (2009). Landslide hazards triggered by the 2008 Wenchuan earthquake, Sichuan China. Landslides, *6*(2), 139–152.

Zhang, Y., Chen, G., Zheng, L., & Li, Y. (2012). Numerical analysis of the largest landslide induced by the Wenchuan earthquake, May 12, 2008 using DDA. In *International Symposium on Earthquake-induced Landslides*. Kiryu, Japan.

Zhang, Y., Chen, G., Zheng, L., Li, Y., & Wu, J. (2013). Effects of near-fault seismic loadings on run-out of large-scale landslide: A case study. *Engineering Geology, 166,* 216–236.

Zhu, W., Zhang, Q., & Jing, L. (1999). Stability analysis of the shiplock slopes of the Three Gorges project by three-dimensional FEM and DEM techniques. In B. Amadei (Ed.), *3rd International Conference on Analysis of Discontinuous Deformation—From Theory to Practice (ICADD-3)* (pp. 263–272). Vail, CO: American Rock Mechanics Association, Alexandria.

Chapter 6
A Case Study of Earthquake-Induced Landslide

Abstract This chapter presents a case study to verify the proposed new methods from slope stability analysis to landslide run-out analysis. The Daguangbao landslide, the largest scale landslide induced by the 2008 Sichuan earthquake, is analyzed using a numerical simulation program DDA as well as FLAC^{3D}. The results show that the vertical component of seismic loading may play an important role in both stability analysis and run-out analysis, as larger tension failure and trampoline effects may be induced by the vertical seismic force, which has generally been ignored up to now.

Keywords Case study · Daguangbao landslide · FLAC^{3D} · DDA

6.1 Introduction

As previous state, a large amount of landslides can be induced by a strong earthquake and cause very serious property damage and human casualties. This phenomenon was recorded at least as early as in ancient China dated back to 1789 BCE (3792 years ago) and in ancient Greece 2385 years ago (Keefer 2002). There have been many reports about very serious damages caused by the earthquake-induced landslides for the last few decades, especially after a series of disastrous earthquake events occurred in recent years. For example, 9272 landslides induced by the 1999 Chi-Chi earthquake (Ms = 7.6) caused 2400 deaths, more than 8000 casualties and over 10 billion US$ of economic loss in Taiwan (Chang et al. 2005). Thirty percentage of the total fatalities (officially 87,350) had been victims of co-seismic landslides due to the 2005 Kashmir earthquake (Ms = 7.6) (Havenith and Boureau 2010). Less than three years later, the 2008 Wenchuan earthquake shock the Sichuan province and induced as many as 60,104 landslides (Gorum et al. 2011), which directly caused more than 20,000 deaths (Yin et al. 2009), a quarter of the total deaths, and over one-third of the total lost caused by the earthquake-induced landslides. In spite of their geomorphic and economic significance, earthquake-induced landslides are not well understood. There are two important issues for

Y. Zhang, *Earthquake-Induced Landslides*, Springer Natural Hazards,
DOI 10.1007/978-981-10-2935-6_6

earthquake-induced landslide. One is the slope fails or not under seismic loadings, say, stability analysis. The other is where and how far it will go once the failure occurs, say, run-out analysis. This chapter focuses on these two issues for a case of earthquake-induced landslide.

For the stability analysis, several applications of continuous modeling to earth structures have been developed and published; Kramer (1996) provided a good summary of various methods and their associated studies. Seed (1973) analyzed the failures of the Upper and Lower San Fernando dams during the 1971 San Fernando earthquake (M 6.6) by using a finite element model to estimate the strain potential at each node based on cyclic laboratory shear tests of soil samples. Lee (1974) and Serff (1976) used the strain-potential method to model the reduction in the stiffness of soils and thus the permanent slope deformation. More recently, nonlinear inelastic soil models have been developed and implemented in 2-D and 3-D models (e.g., Prevost 1981; Griffiths and Prevost 1988; Elgamal et al. 1990; Taiebat et al. 2011). In addition, Zheng et al. (2009) and Latha and Garaga (2010) studied the seismic slope stability by using FDM. FLAC^{3D}, a finite difference method, is used in this chapter to evaluate the initiation of earthquake-induced landslide.

For the run-out analysis, many numerical methods now exist to investigate the dynamic process of landslide (Savage and Hutter 1989; Chen and Lee 2000; Denlinger and Iverson 2001; Crosta and Agliardi 2003; Crosta and Frattini 2003; McDougall and Hungr 2004; Crosta et al. 2005, 2007; Chen et al. 2006). These methods are usually based on continuum mechanics and assume that the avalanche thickness is very much smaller than its extent parallel to the bed, i.e., thin layer depth-averaged models. These models can take account accurately detailed topography effects, shown to be significant, with a reasonable computational time, making it possible to perform sensitivity studies of the parameter used in the model. They can provide effective properties that make it possible to roughly reproduce the deposit shape but also the dynamic as shown in Favreau et al. (2010), and Moretti et al. (2012) for examples. However, conventional continuum approaching models, which neglect the contact between rocks, make it impossible to trace the position of individual rock during a landslide. In contrast, discontinuum numerical simulation methods are powerful tools in simulation of failure and run-out process of rock avalanche controlled by weakness surface. DEM (Cundall 1971) and DDA (Shi and Goodman 1985, 1989) are two of the most commonly used methods. Differences between DEM and DDA are stated in Chap. 2. Compare to DEM, DDA has a simpler and more straightforward physical meaning (Wu 2003). Therefore, DDA is used herein to evaluate the run-out of earthquake-induced landslide.

The studied case is the Daguangbao landslide induced by the 2008 Wenchuan earthquake. The Daguangbao landslide, with an estimated influenced area of about 7.3–10 million m^2 and a volume of 750–840 million m^3, is the biggest one induced by the 2008 Wenchuan earthquake which is not only the largest known landslide in China so far, but also the one of the world's known largest landslides. Immediately after the sliding, several times, inspections have been made by Huang's group of Chengdu University of Technology, China. Their papers (Huang et al. 2008, 2009, 2012) gave a general introduction of the Daguangbao landslide focused on the characteristics, failure mechanism, and geological property. After that, several post-earthquake field investigations have been carried out by them, and more deeply studies on characteristics and failure mechanism have been presented (Huang et al. 2009, 2012). In addition, remote sensing research on the Daguangbao landslide has been carried out by Yin et al. (2011). These researches provide the best information available. Field investigations showed that tensile failure occurred at the back edge that due to the effect of a large vertical seismic motion that occurred in the meizoseismal area during the earthquake. In order to investigate the effects of seismic force on the Daguangbao landslide, FLAC^{3D} and DDA are used to simulate the initiation and run-out, respectively.

Including this introduction, the chapter consists of six sections. Section 6.2 briefly describes the background of the Daguangbao landslide. Section 6.3 outlines the material properties and seismic loadings. Sections 6.4 and 6.5 analyze the stability and run-out processes of the Daguangbao landslide using FLAC^{3D} and DDA, respectively. Finally, conclusions are drawn in Sect. 6.6.

6.2 Background Information

6.2.1 The Wenchuan Earthquake

The earthquake had a magnitude of M_w = 7.9, occurred in Sichuan Province, China, at 14:28 CST on May 12, 2008. The epicenter is located in Yingxiu town (30.986°N, 103.364°E), Wenchuan County. The focal depth is about 12 km according to the report by China Earthquake Administration (CEA). The earthquake occurred along the Longmenshan fault (LMSF) zone at eastern margin of the Tibetan Plateau, adjacent to the Sichuan Basin as shown in Fig. 6.1 (Gorum et al. 2011). The fault belt consists of a series of active parallel thrusts, among which, the Wenchuan–Maowen fault (F1 in Fig. 6.1), Yingxiu–Beichuan fault (F2 in Fig. 6.1), and Pengxian–Guanxian fault (F3 in Fig. 6.1) are considered to be seismogenic. The LMSF locates in a north–south zone of high topographical and geophysical gradients between the western Tibet Plateau and the eastern Yangzi Platform. Seismic activities concentrated on its mid-fracture (known as Yingxiu–Beichuan fracture). Starting from Yingxiu, the rupture propagated unilaterally toward the northeast at an average speed of 3.1 km/s, generating a 300-km- and a 100-km-long surface rupture along the Yingxiu–Beichuan and Pengguan faults, respectively (Huang et al. 2011). The duration was as long as 120 s, and the maximum displacement amounted to 9 m. Official figures, released by China News www.chinanews.com, on July 21, 2008 12:00 CST show that 69,197 were confirmed dead, 374,176 injured, and 18,222 listed as missing. The earthquake destroyed 5,362,500 and seriously damaged 21,426,600 houses, left about 4.8 million people homeless

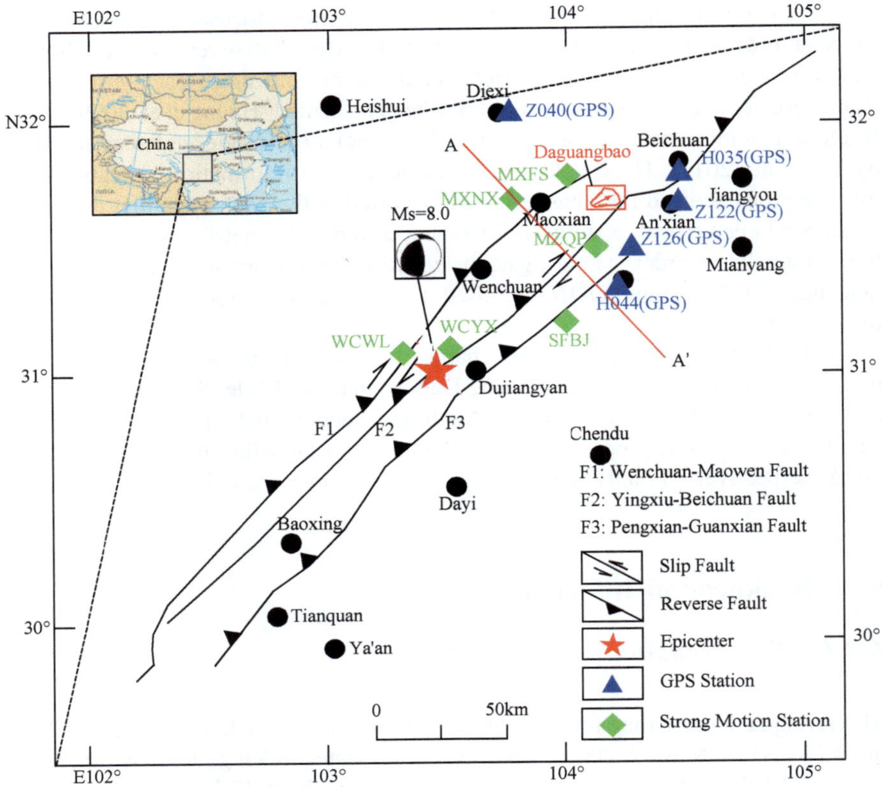

Fig. 6.1 Locations of the Daguangbao landslide, GPS stations, and strong motion stations

(Cui et al. 2009; Tang et al. 2011). Approximately 15 million people lived in the affected area. It was the deadliest earthquake to hit China since the 1976 Tangshan earthquake, which killed at least 240,000 people.

6.2.2 *The Daguangbao Landslide*

6.2.2.1 Geological Setting

The Daguangbao landslide is located in the hanging wall only 6.5 km away from the Yingxiu–Beichuan fault. Figure 6.2 shows the pre-earthquake 3-D topography model of the Daguangbao landslide which has very distinctive scarps and flanks. The pre-earthquake topography of the landslide area indicates that the elevation descended from the west to the east, with the highest points of 3047 m above sea level (a.s.l.) at the peak of the Daguangbao point and the lowest point of 1450 m above sea

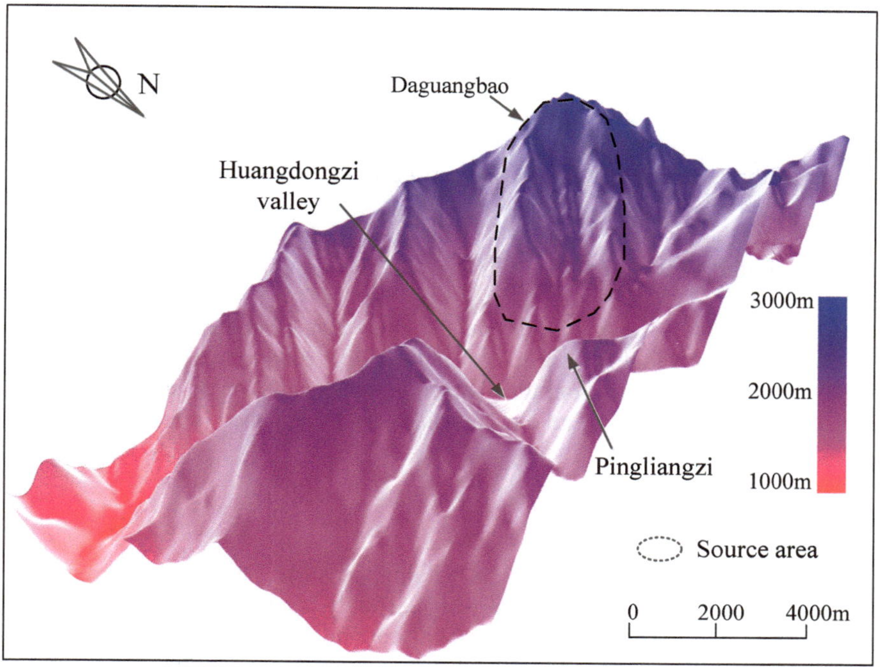

Fig. 6.2 Pre-earthquake 3-D topography model of the Daguangbao area (The data set is provided by International Scientific & Technical Data Mirror Site, Computer Network Information Center, Chinese Academy of Sciences)

level at the bottom of the Huangdongzi valley. The altitude difference is about 1600 m over a horizontal distance of 3 km (Huang et al. 2012) (Fig. 6.2). Along the main sliding direction, the terrain of the landslide can be classified into three sections. The first section is the Daguangbao peak that ridge altitude varies from 3047 to 2700 m a.s.l. and the slope angle ranges from 50° to 60°. The second section is the middle part of the slope with an average slope angle of 30° and an altitude difference of 700 m, decreasing from 2700 to 2000 m a.s.l. And the third section varies from 2000 to 1500 m a.s.l. with a slope angle ranging from 40° to 50°.

Figure 6.3 shows a geologic map of the Daguangbao area. The strata in the landslide area are mainly composed of carbonates. Due to the thrust–nappe effect of the LMSF belt and the long-term denudation, the lithology can be classified into the following: (1) the Feixianguan group of the Triassic system (T_f^2), which consists of the violet siltstone, silty mudstone with a small amount of shell, and micritic limestone; (2) the Liangshan group (P_l), Yangxin group (P_y), Longtan group (P_{lt}), and Wujiaping group (P_w) of the Permian system, which are dominantly composed of the gray-to-dark-gray and medium-to-thick layered crystalline and micritic limestone with flint, the bioclastic rocks; (3) the Zongchang group of the

Carboniferous system (C_z^1), which consists of violet red sandy mudstone; (4) the Shawozi group of the Devonian system (D_s), which contains dolomite with phosphate mineral rocks; and the (5) Dengyin group of the Sinian system (Z_d^1, Z_d^2, and Z_d^3), which is composed of limestone, interbeds red mudstone, and dolomitic rocks (Huang et al. 2012) (see Fig. 6.3). A more detailed description about the geological setting of the Daguangbao slope zone can be found in Huang et al. (2012).

6.2.2.2 Failure History

Due to the scarps in first section and the smooth and moderate slope in second section, it is estimated that some historical landslides might have occurred along the strike of the strata. In addition, the distinctive flanks and convex slope surface also indicate that there might be an old landslide which happened in history (Fig. 6.2).

6.2.2.3 Failure Type

The Daguangbao landslide is a typical bedding landslide. The southern boundary of the landslide is formed by the strata, which actually is a part of the exposed sliding surface. The sliding surface was exposed discontinuously on the slope surface with the ranging from 1900 to 2400 m a.s.l. It trends 80°N–88°E, dipping at 35°–38°, toward the NW, along the bedding. Apparent slickensides can be observed on the exposed sliding surface, having a pitch of 30°, which reflects that the angle between the sliding direction and the strike of sliding surface is acute. Thus, the sliding direction dominantly trends along the strike of the sliding surface (Huang et al. 2012).

Figure 6.4 gives a pre- and post-earthquake 3-D topographies, and Fig. 6.5a shows an *Advanced Land Observing Satellite* (ALOS) image of the landslide photographed immediately on June 4, 2008 after the May 12th Wenchuan earthquake. The image has a resolution of 2.5 m and was taken by the Panchromatic Remote sensing Instrument for Stereo Mapping, which is comprised of three sets of optical systems to measure precise land elevation (JAXA: http://www.jaxa.jp/projects/sat/alos/index_e.html). The scale of the landslide can be appreciated with the satellite photos presented in Fig. 6.5b and c which provide a bird's eye view and a northeast view of the landslide, respectively. The sliding mass of the Daguangbao landslide formed a 3.2-km-long and 2.2-km-width deposition zone, with a maximum thickness of about 550 m. The sliding mass crossed the Huangdongzi valley and ran up the opposite slope with a height of more than 500 m. The landslide travelled such a long distance and climbed over the Pingliangzi. There must be a rapid speed in the run-out progress. Figure 6.6 shows the profile of the Daguangbao landslide which was constructed based upon the contour maps of the area before

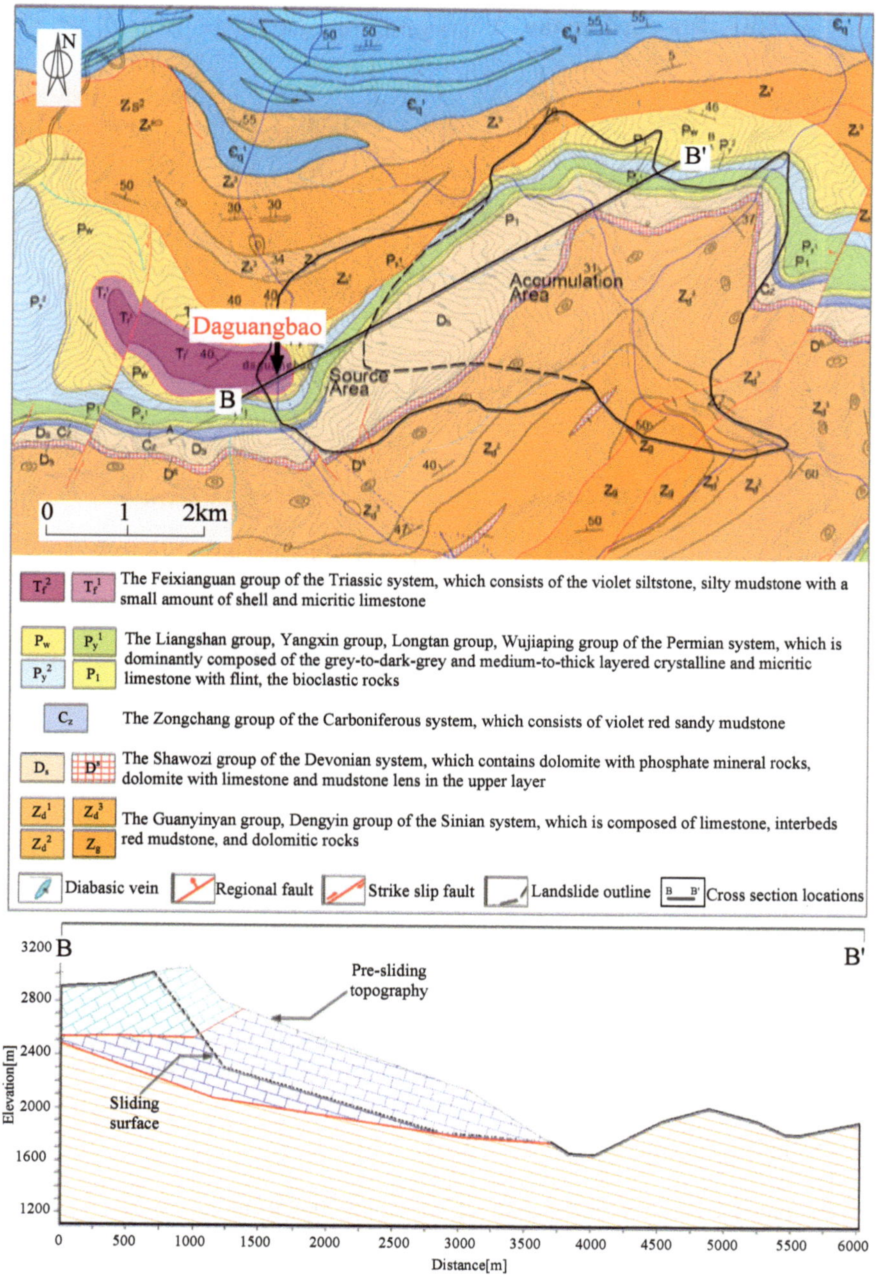

Fig. 6.3 Geology near the Daguangbao landslide (modified from Huang et al. 2012) and B-B′ cross section

Fig. 6.4 Pre- and post-earthquake 3-D topographies of the Daguangbao landslide

and after the Wenchuan earthquake (Fig. 6.4). The extent of the damage caused by the Daguangbao landslide is reflected in the following statistics (Huang et al. 2008, 2009, 2012; Yin et al. 2011; Zhang et al. 2011):

1. The affected area covered 7.3–10 km^2;
2. The accumulation body width is 2.2 km;
3. Estimated volume of collapsed rock mass is 750–840 million m^3;
4. The sliding surface is more than 1 km in the source area;
5. The sliding mass travelled about 4.5 km;
6. Formed a landslide dam more than 600 m high.

Fig. 6.5 **a** Post-earthquake air-photo of the Daguangbao area photographed on June 4, 2008 [Modified from the Japan Aerospace Exploration Agency (JAXA)]; **b** Satellite image of the Daguangbao landslide in a bird's eye view (Modified from Google); and **c** Satellite image of the Daguangbao landslide in a northeast view (Modified from Google)

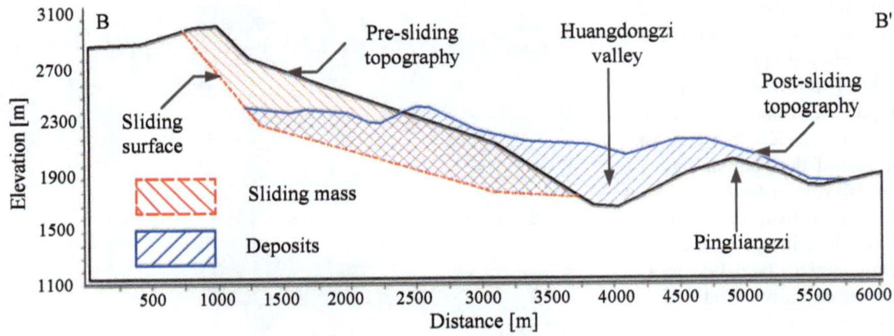

Fig. 6.6 Cross section of the Daguangbao landslide before and after the Wenchuan earthquake. The cross section is the B-B′ section in Fig. 6.5b, c

6.3 Material Properties and Ground Motion

6.3.1 Material Properties

The Daguangbao landslide is so huge that the size effect must be considered. Size effect means the deposits of a natural rock avalanche with a volume larger than approximately 10^6–10^7 m^3 typically extend much farther than those of smaller avalanches and extend much farther than the deposits simulated by a friction model (Scheidegger 1973; Hsu 1975; Davies and McSaveney 2009; Erismann and Abele 2001). The long run-out is thus not expected to relate to the friction coefficient measured in the laboratory. To account for this discrepancy, experience equations based on Hoek–Brown failure criterion, which size effect can be considered, are used to back-calculate the material strength (Hoek 1990; Hoek et al. 2002).

Hoek and Brown (1988), Hoek et al. (2002) proposed an empirical failure criterion which developed through curve fitting of triaxial test data suited for intact rock and jointed rock masses. The Hoek–Brown criterion is one of the few non-linear failure criteria widely accepted and used. The latest version of Hoek–Brown failure criterion, solid line in Fig. 6.7, is used.

$$\sigma_1' = \sigma_3' + \sigma_{ci}\left(m_b\frac{\sigma_3'}{\sigma_{ci}} + s\right)^a \tag{6.1}$$

where σ_1' and σ_3' are the maximum and minimum effective stresses at failure, respectively; m_b is the value of the material constant m_i and is given by $m_b = m_i \exp[(GSI - 100)/(28 - 14D)]$; s and a are constants for the rock mass given by the following relationships: $s = \exp[(GSI - 100)/(9 - 3D)]$ and $a = 1/2 + 1/6\left(e^{-GSI/15} - e^{-20/3}\right)$, D is a factor which depends upon the degree of disturbance to which the rock mass has been subjected by blast damage and stress

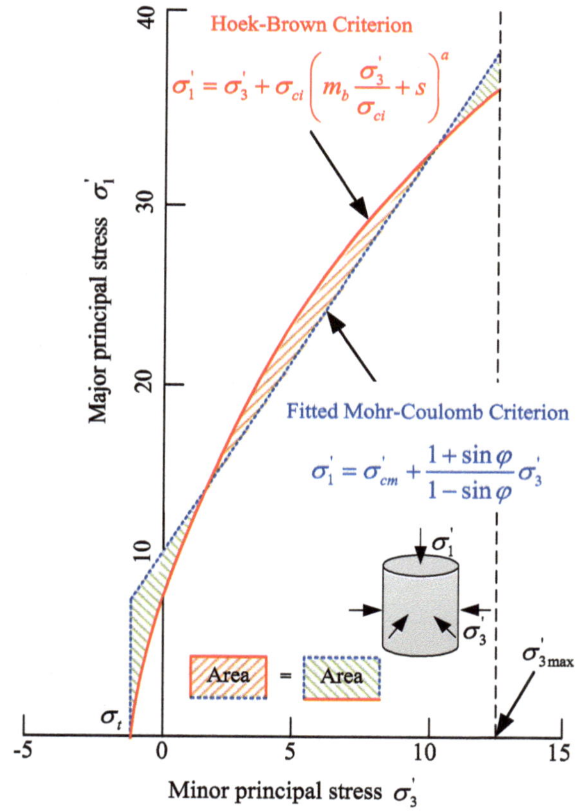

Fig. 6.7 Relationships between major and minor principal stresses for Hoek–Brown and fitted Mohr–Coulomb criteria

relaxation. It varies from 0 for undisturbed in situ rock masses to 1 for very disturbed rock masses.

As the nonlinear shear strength from the Hoek–Brown failure criterion cannot be directly used in DDA, it is necessary to determine equivalent friction angle and cohesive strength used in linear Mohr–Coulomb criterion. This is done by fitting an average linear relationship to the curve generated by solving Eq. 6.1 (point line in Fig. 6.7). The fitting process involves balancing the areas above and below the Mohr–Coulomb plot. The fitted Mohr–Coulomb criterion is given by

$$\sigma_1' = \sigma_{cm}' + \frac{1 + \sin\varphi}{1 - \sin\varphi}\sigma_3' \tag{6.2}$$

where σ_{cm}' is the rock mass strength, defined by

$$\sigma_{cm}' = \sigma_{ci}\frac{[m_b + 4s - a(m_b - 8s)]\left(\frac{m_b}{4} + s\right)^{a-1}}{2(1+a)(2+a)} \tag{6.3}$$

The equivalent friction angle and cohesive strength are:

$$\varphi = \arcsin\left[\frac{6am_b(s+m_b\frac{\sigma'_{3\max}}{\sigma_{ci}})^{a-1}}{2(1+a)(2+a)+6am_b(s+m_b\frac{\sigma'_{3\max}}{\sigma_{ci}})^{a-1}}\right] \tag{6.4}$$

$$c = \frac{\sigma_{ci}\left[(1+2a)s+(1-a)m_b\frac{\sigma'_{3\max}}{\sigma_{ci}}\right](s+m_b\frac{\sigma'_{3\max}}{\sigma_{ci}})^{a-1}}{\sqrt{(1+a)(2+a)\left[(1+a)(2+a)+6am_b(s+m_b\frac{\sigma'_{3\max}}{\sigma_{ci}})^{a-1}\right]}} \tag{6.5}$$

Obviously, the results in the above equations note that the angle of friction φ and cohesive strength c are the functions of the maximum of minor principal stress $\sigma'_{3\max}$. In order to estimate the angle of friction φ and cohesive strength c, appropriate value of $\sigma'_{3\max}$ should be determined firstly for each individual case. Hoek and Brown gave a fitted equation for a wide range of slope geometries and rock mass properties:

$$\sigma'_{3\max} = 0.72\sigma'_{cm}\left(\frac{\sigma'_{cm}}{\gamma H}\right)^{-0.91} \tag{6.6}$$

where γ is the unit weight of the soil and H is the height of slope.

For different $\sigma'_{3\max}$, there are different fitted Mohr–Coulomb criterion (Fig. 6.8a). Take two slopes with same material parameters but different height H as examples, the solid line in Fig. 6.8a indicates the original nonlinear Hoek–Brown failure criterion of the slope material, while the point line and dash line indicate the fitted linear Mohr–Coulomb criterion of soil materials of small-scale and big-scale slopes, respectively. Figure 6.8b shows the changed criteria from $\sigma'_3 - \sigma'_1$ coordinate system to $\sigma - \tau$ coordinate system.

Experience values of input index for Hoek–Brown criterion are listed in Table 6.1, in which σ_{ci} *GSI* and m_i are decided based on Hoek and Brown (1997); *D* is decided based on Hoek et al. (2002); γ is estimated based on engineering practice experience. It should be noted that it contains some associated uncertainties in the estimation of material properties, e.g., *GSI*, *m,* and *D*.

Table 6.2 lists some output strength parameters. Table 6.3 shows the analytical conditions and material properties.

6.3.2 Ground Motion

For the present study, there are two problems needed to be addressed regarding the adoption of input ground motions. The first problem is selecting appropriate strong ground motions from the host of records as input. The other one is baseline

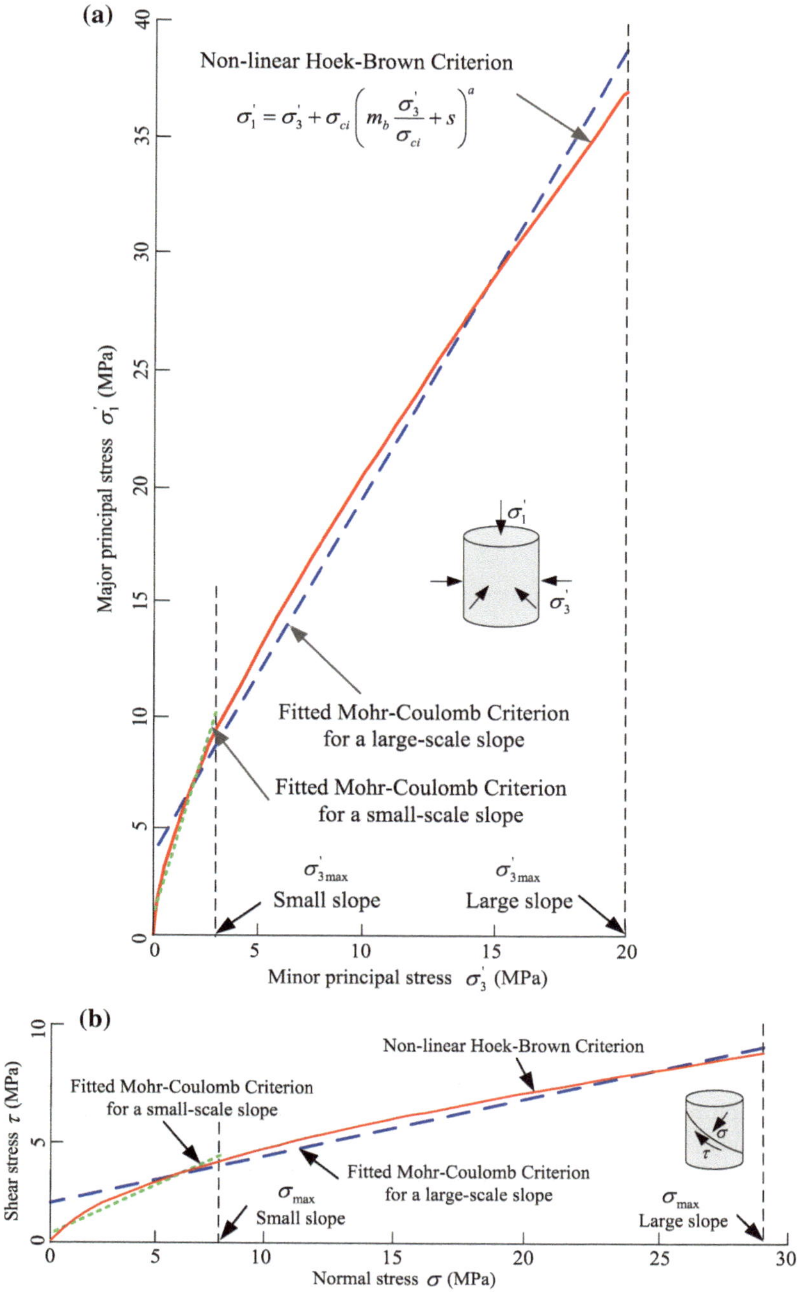

Fig. 6.8 Fitted Mohr–Coulomb criteria for different scale slope with the same material in **a** $\sigma_3' - \sigma_1'$ and **b** $\sigma - \tau$ coordinate system

Table 6.1 Input index of the Hoek–Brown experience equations

	σ_{ci} (MPa)	GSI	m_i	γ (kN/m^3)	D
Material 1	43.8	40	12	25	1
Material 2	87.2	40	9	26	1
Material 3	87.2	70	7	26	1

Table 6.2 Output index of back calculation for rock strength parameters using the Hoek–Brown experience equations

	σ_t (kPa)	c (MPa)	φ (degree)	E (GPa)
Material 1	12	1.276	10.8	1.86
Material 2	32	1.576	12.2	2.63
Material 3	556	4.052	23.53	14.76

Table 6.3 Material properties of the Daguangbao landslide in FLAC^{3D} and DDA

	Material 1	Material 2	Materia 3	
			FLAC^{3D}	DDA
Density (ρ): g/cm^3	2.5	2.6	2.6	260,000
Unit weight of rock (γ): kN/m^3	25	26	26	0
Elastic modulus (E): GPa	1.86	2.63	14.76	
Poisson's ratio (ν)	0.2	0.2	0.1	
Friction angle of discontinuities (φ): °	10.8	12.18	23.53	
Cohesion of discontinuities (c): MPa	1.276	1.576	4.052	
Tensile strength of discontinuities (σ_t): kPa	12	32	556	

correction of the selected near-fault motions records that with a large co-seismic displacement that different from general far-fault acceleration records.

6.3.2.1 Selection of Input Ground Motion

At time of the 2008 Wenchuan earthquake, China's digital accelerograph network, under the auspices of China Earthquake Administration, collected a rich set of acceleration records. Two criteria for choosing acceleration records are considered in this study. The first criteria is that the accelerograph station should be located as close to the Daguangbao landslide as possible. The second one is that the selected seismic loadings should reflect the characteristic of the earthquake as far as possible, e.g., the integrated residual displacement from acceleration time histories should be close to the real ground co-seismic displacement. Several candidate accelerograph stations are located close to the Daguangbao landslide as shown in Fig. 6.1. According to the latest studies on the surface co-seismic displacement of

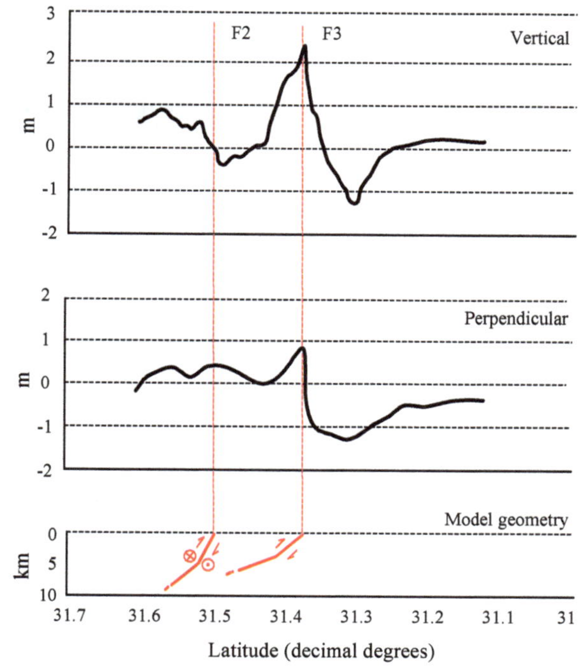

Fig. 6.9 Vertical and perpendicular components across-the-strike of the co-seismic surface displacement measured at transect locations AA′ (modified from de Michele et al. 2010)

the Wenchuan earthquake (de Michele et al. 2010), Pengxian–Guanxian fault (F3 in Fig. 6.1) dominates the surface displacement trend in the study area although the Yingxiu–Beichuan fault is the main shock fault (F2 in Fig. 6.1). Figure 6.9 shows the vertical and perpendicular components across-the-strike of the co-seismic surface displacement measured at transect locations AA′ (Fig. 6.1). Although the MXFS and MZQP stations have almost the same displacements from the studied area, MZQP station is more close to main fault and it brings the ampler earthquake information than MXFS station. Hence, the MZQP station records are adopted as input source for analysis, and these acceleration records are depicted in Fig. 6.10. Each set of seismic data is a 100 s acceleration record, comprising 20 s of pre-event data and 80 s of the earthquake data. The peak ground accelerations from the station are −0.824 g for E–W component, 0.803 g for N–S component, and 0.623 g for U–D component, respectively.

6.3.2.2 Baseline Correction

Baseline correction is usually conducted to eliminate the integration errors for time domain seismic analysis. For near-fault seismic records with a large tilt of the ground, linear correction, i.e., a first-order polynomial, from the velocity is a useful method (Iwan et al. 1985). A composite baseline with two segments has been

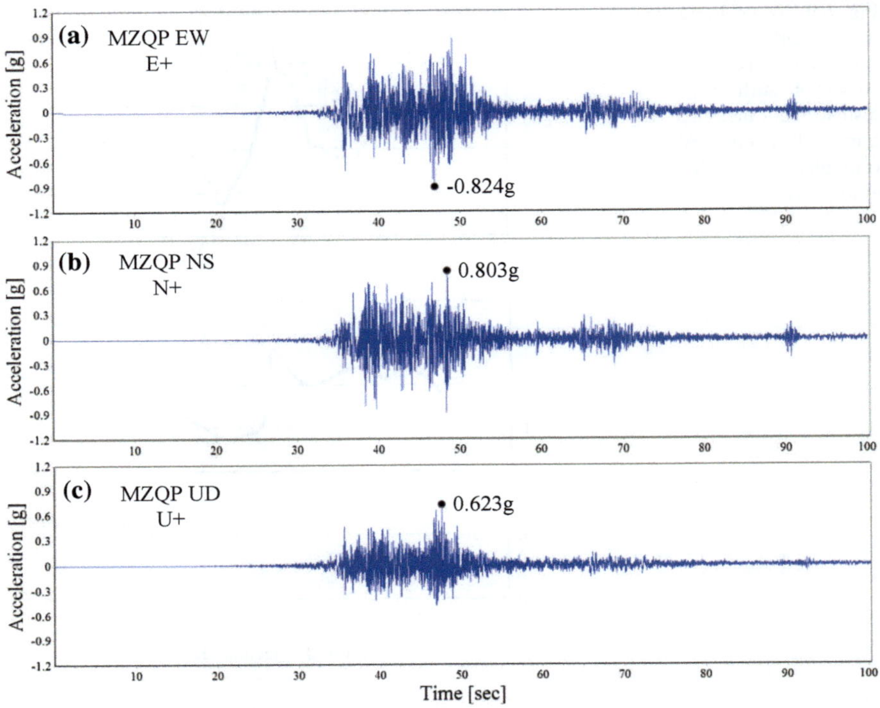

Fig. 6.10 Ground accelerations records of strong motion MZQP of the Wenchuan earthquake. **a**, **b**, and **c** are components of east–west (E–W), north–south (N–S), and up–down (U–D), respectively. Symbol + means the positive direction

successfully applied in baseline correction of near-fault acceleration records of the 1998 Rueyli earthquake, Taiwan, and 1999 Chi-Chi earthquake (Boore 2001; Wu et al. 2009; Wu 2010; Wu and Tsai 2011). Because of the complexity of acceleration records from the Wenchuan earthquake, an extended baseline correction approach called multi-segment linear baseline correction is presented in this chapter. Figure 6.11 shows the flowchart of the multi-segment linear baseline correction and process of a four-segment example.

Two linear baselines for two horizontal records and three linear baselines for vertical records are used for baseline correction in this paper. As an illustration, the baseline correction progress for horizontal record is depicted in Fig. 6.12 in which two linear baselines are first obtained, one for the first 20 s pre-event and the other for the last 20 s of the velocity history which is obtained by once integration from acceleration record. These lines are extended until they meet and that gives a composite baseline. The displacement then becomes normal without non-convergence.

The permanent ground displacements, obtained by double integrating the seismic accelerations, must be identical to the ground movements measured by the GPS at the same position. However, the locations of the GPS stations are not entirely

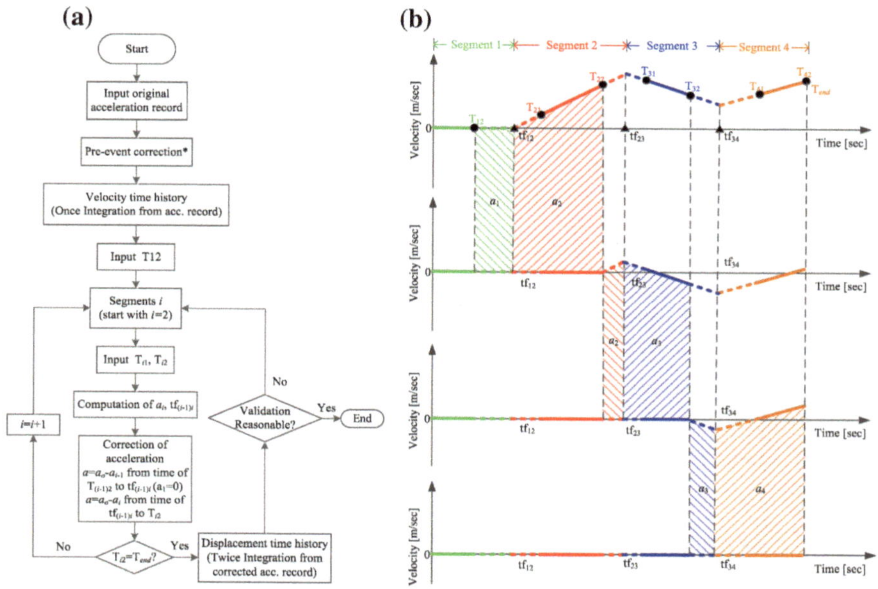

Fig. 6.11 Multi-segments linear baseline correction approach. **a** Flowchart of the baseline correction program. **b** Baseline correction process of an example with four segment

consistent with the strong motion stations in China. Table 6.4 represents the co-seismic displacements that were measured during the Wenchuan earthquake by the GPS stations (Z126, Z122, H035, and H044) that are near the Daguangbao landslide and located in the same southeastern foothill geologic zone. The baseline corrections are applied by trial and error until the co-seismic displacement is close to the data that were measured at the neighboring GPS stations. Figure 6.13 shows the comparisons of corrected co-seismic displacement histories with the data from GPS stations. This provides some credence to the baseline correction used.

6.4 Stability Analysis by the FLAC^{3D}

6.4.1 The Finite Difference Method

The finite difference method code, FLAC^{3D} (Fast Lagrangian Analysis of Continua in 3-Dimension) (Itasca 2007), is used to carry out the computation. The calculation of FLAC^{3D} is based on the explicit finite difference scheme to solve the full equations of motion, using lumped grid point masses derived from the real density of surrounding zones. For given element shape functions, the set of algebraic equations solved by FLAC^{3D} is identical to that solved with the finite element

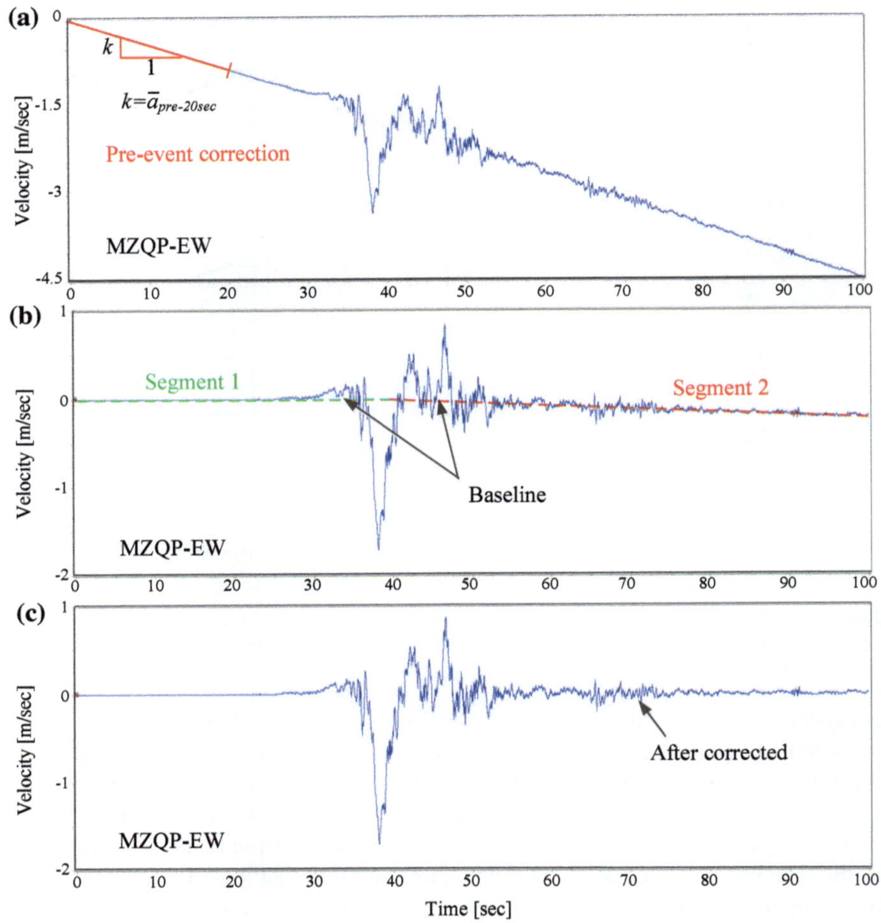

Fig. 6.12 The baseline correction scheme and ground velocity history of MZQP E–W component after baseline removal

Table 6.4 Permanent displacement from GPS stations near the Daguangbao landslide

Station	Longitude (°)	Latitude (°)	E–W displacement (cm)	N–S displacement (cm)	U–D displacement (cm)
Z040	103.68	32.04	31.0	−3.4	–
Z126	104.25	31.51	−122.1	37.9	−20.4
H044	104.19	31.35	−98.3	39.7	−12.3
H035	104.44	31.80	−237.9	48.1	−67.5
Z122	104.45	31.69	−94.5	43.9	−46.6

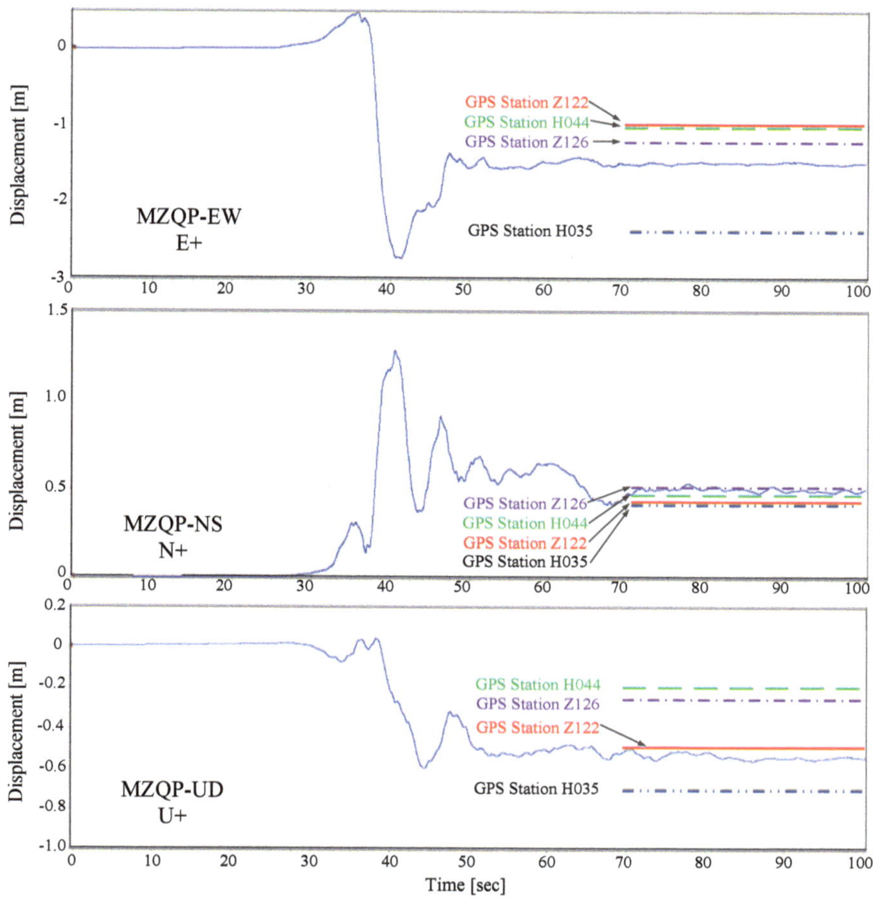

Fig. 6.13 Comparisons of corrected co-seismic displacement histories with data from GPS stations

method. However, in FLAC^{3D}, this set of equations is solved using dynamic relaxation, an explicit, time-marching procedure in which the full dynamic equations of motion are integrated step by step. FDM has the advantages of efficient storage capacity, fast calculation, as well as good convergence.

The convergence criterion for FLAC^{3D} is the nodal unbalanced force, the sum of forces acting on a node from its neighboring elements (Dawson et al. 1999). If a node is in equilibrium, these forces should sum to zero. For this study, the maximal unbalanced force of all nodes is normalized by the gravitational body force acting on that node. A simulation is considered to have converged when the normalized unbalanced force of every node in the mesh is less than 10^{-5}.

6.4.2 FOS and Strength Reduction Technique

A generally accepted definition of FoS is that the ratio of the available strength of soil material to that required to maintain equilibrium. Based on this definition, the current approach is to use a shear strength reduction technique in which reduced shear strength parameters c' and φ', given by (Zienkiewicz et al. 1975)

$$c' = \frac{c}{\mathrm{SRF}},\ \varphi' = \arctan\frac{\tan\varphi}{\mathrm{SRF}} \tag{6.7}$$

are used here, where c and φ are the actual values of soil material, cohesion and friction angle, respectively, c' and φ' are the relevant parameters required to maintain the limit equilibrium, and SRF is a strength reduction factor. To obtain the correct factor of safety, it is essential to trace the SRF gradually until the reduced strength parameters c' and φ' bring the slope to a limit equilibrium state. In this process, the SRF is assumed to apply equally to both c and φ. When the slope reaches the limit state, the FoS = SRF. The strength reduction technique is identical to that used in the LEM and LAM. The calculation of FoS is automatically carried out by the FLAC^{3D} program.

6.4.3 Numerical Simulations

6.4.3.1 Cases State

There are five cases studied in this paper. Before analyzing the influence of the Wenchuan earthquake on the Daguangbao landslide, the stability in its natural state should be analyzed first. In order to analyze the effect of the vertical seismic force on the failure mechanism of the slope, two cases in pseudo-static situation and two cases in dynamic situation, with and without the vertical seismic force, respectively, are considered. Hence, the cases studied in here are drawn following:

Case 1 Static case,
Case 2 Pseudo-static only-horizontal case,
Case 3 Pseudo-static horizontal-and-vertical case,
Case 4 Dynamic only-horizontal case,
Case 5 Dynamic horizontal-and-vertical case.

In case 2 and case 3, 0.3 and 0.5 times of the PGA are considered. In addition, for case 3, the acting direction of vertical seismic force is taken as upward and downward directions.

6.4.3.2 Models

In order to improve the calculation efficiency, two models with different mesh size for static and dynamic cases are used in this paper. According to the geological characteristics of the Daguangbao area and the post-earthquake field investigations, static model with mesh size of 50 m is shown in Fig. 6.14. Field investigation shows that the rock mass below the sliding surface is relatively intact, but the upper rock mass is strongly weathered and broken. To facilitate the modeling calculations for geometric mechanics, the geological model is generalized as three materials.

Different from static situation, there are several important aspects should be considered while preparing a FLAC^{3D} model for dynamic analysis: (1) boundary conditions, (2) mesh size, and (3) mechanical damping.

6.4.3.3 Boundary Conditions

As for the dynamic analysis, in order to make the fluctuations meet the radiation phenomena in the original continuous media, the artificial boundary must be set up on the actual boundary. In this paper, the viscous boundary conditions in FLAC^{3D} are used to ensure the precision of simulation results (Fig. 6.15).

6.4.3.4 Mesh Size

In dynamic analysis, the mesh size must meet the requirement of the wave propagation in the simulation model precisely. The appropriate mesh size can be chosen based on the following relationship (Itasca 2007):

$$\Delta l \leq \left(\frac{1}{8} - \frac{1}{10}\right)\lambda \tag{6.8}$$

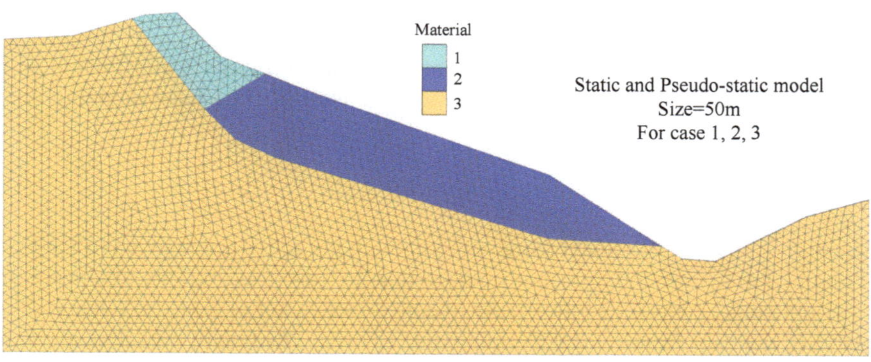

Fig. 6.14 FLAC^{3D} static and pseudo-static model of the Daguangbao landslide

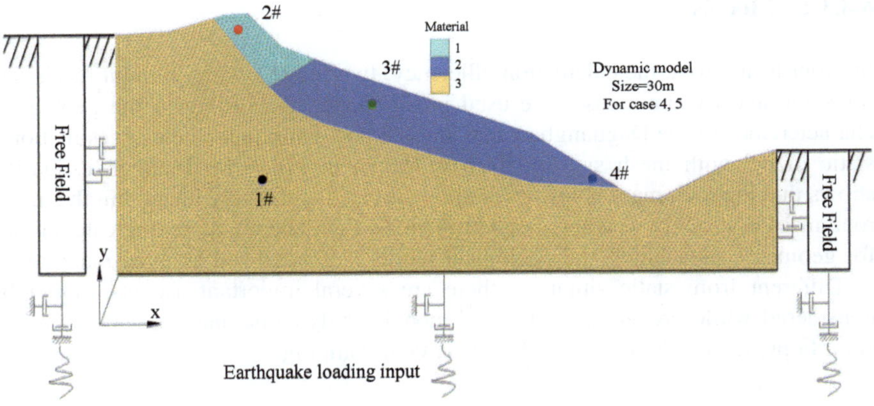

Fig. 6.15 FLAC^{3D} dynamic model of the Daguangbao landslide

where Δl is the maximum mesh size of simulation model, λ is the shortest wavelength of incident wave, and it can be defined by needless to say

$$\lambda = \frac{c_s}{f} \tag{6.9}$$

$$c_s = \sqrt{\frac{E}{2\rho(1+v)}} \tag{6.10}$$

where c_s is the minimum wave velocity of seismic wave, f is the maximum frequency of incident wave; E, ρ, and v are the elastic modulus, density, and Poisson's ratio, respectively. The length of mesh size for dynamic analysis is 10 m, which can meet the accuracy requirements.

6.4.3.5 Mechanical Damping

For a dynamic analysis, FLAC^{3D} program provides several mechanical damping in which local damping is a simple and pragmatic method. The local damping coefficient α_L defined as

$$\alpha_L = \pi D \tag{6.11}$$

where D is fraction of critical damping. Although the actual value given to the local damping has a profound influence on the dynamic wave transmission, if chose from a certain range, it has little influence on the predicted factor of safety in seismic slope stability analysis. Hence, local damping of 0.157 (i.e., fraction of critical damping is 5%) is used in the model as suggested by other studies for these kinds of problems.

The FLAC^{3D} model of the Daguangbao landslide for dynamic analysis is shown in Fig. 6.15, in which four monitoring points are set up.

6.4.3.6 Dynamic Loading

It is known that for a seismic wave, it often contains a wide spectrum, but most of the energy is concentrated in the part with a frequency of less than 10 Hz. In fact, the high-frequency seismic wave does not play an important role on slope failure. Hence it can be filtrated through the filtering method. In this study, frequency more than 10 Hz is filtrated. Figure 6.16 shows the filtered original seismic waves along slope strike and perpendicular to slope, and Fig. 6.17 shows the Fourier spectrums of input seismic wave. The horizontal earthquake wave is the projection combination of the E–W and N–S components of MZQP acceleration records in the main sliding direction (N60°E) (as Eq. 6.12). The inputted vertical earthquake wave is the MZQP acceleration records in U–D direction.

$$a_H = a_{E-W} \cdot \sin 60^\circ + a_{N-S} \cdot \cos 60^\circ \tag{6.12}$$

Figure 6.18 shows the input combined acceleration records. Velocity and displacement time histories can be obtained by first and second integration from

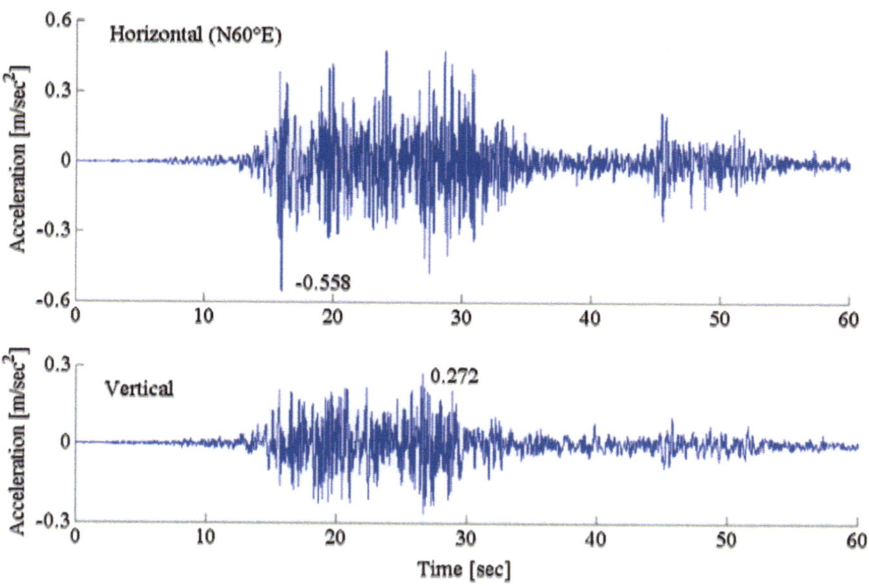

Fig. 6.16 Filtered ground accelerations records of MZQP strong motion station of the Wenchuan earthquake

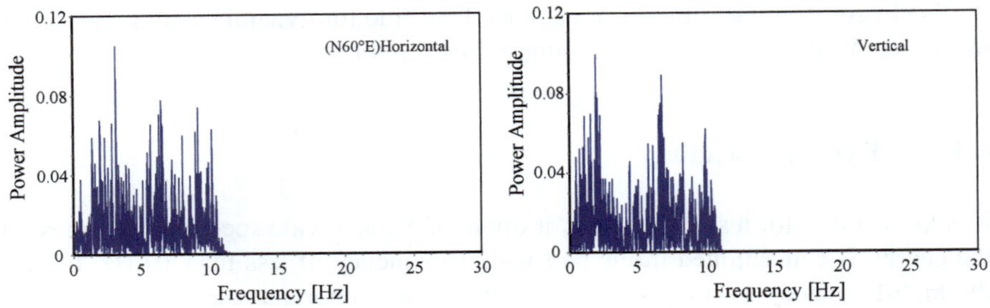

Fig. 6.17 Fourier spectrum of the filtered input seismic wave

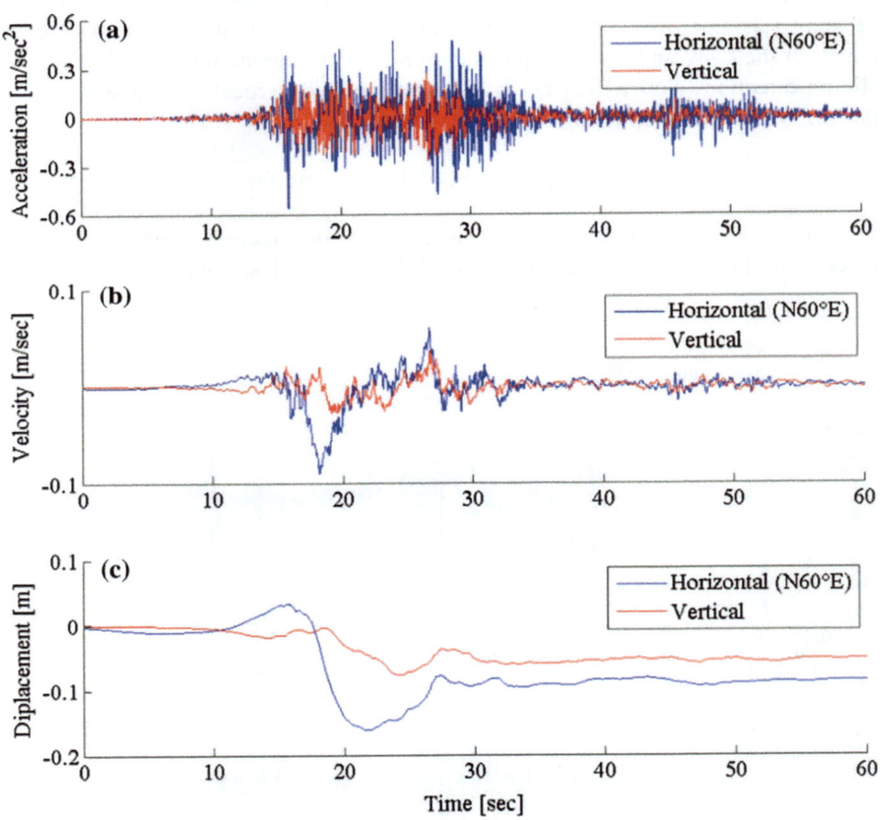

Fig. 6.18 **a** Input post-corrected horizontal and vertical ground acceleration records, **b** velocity, and **c** displacement histories projected to N60°E direction

acceleration record. The duration of earthquake wave is 80 s. In case 4, only the horizontal record is taken into account; but in case 5, both the horizontal and vertical records are considered.

6.4.4 *Results and Discussions of the Stability Analysis*

6.4.4.1 FoS

Table 6.5 lists the factor of safety of the first three cases, static and pseudo-static cases. For case 1, the factor of safety in the natural state of the Daguangbao is 1.16, which indicates that the slope was stable before the Wenchuan earthquake. The factor of safety of the Daguangbao landslide in case 2 and case 3 is smaller than 1.0, which indicates the seismic conditions cause a significant reduction in factor of safety than static situation and take the slope into unstable state.

If take the pseudo-static seismic coefficient as 0.3 times of peak ground acceleration and consider the vertical seismic force acting on the upward direction, the factor of safety of the Daguangbao landslide in case 2 is 0.76 and 0.79 in case 3, which indicates that the vertical seismic force has insignificant influence on factor of safety in pseudo-static situation. Factor of safety from other situations of different pseudo-static seismic coefficients and different vertical acting directions (upward or downward) has the same result. This is identical with the traditional opinion: The vertical seismic force always decreases or increases the resistance force and sliding force at the same time. That is the reason of why the vertical seismic force is always ignored in engineering practice.

6.4.4.2 Tension Failure Area

Figure 6.19a–e show the distribution maps of the tension failure area in various cases. It can be seen that from case 1 to case 3, the range of tension failure becomes larger and larger, especially in the back edge of the slope. From the distribution maps of tension failure in case 4 and case 5, the same distribution trend can be found. Results show that when the vertical seismic force is considered, a larger tension failure area will occur at the head scarp.

6.4.4.3 Permanent Displacements

During simulation, four monitoring points, one stable point 1# under the potential sliding surface and three unstable points 2#, 3#, and 4# above the potential sliding surface, are established and monitored (Fig. 6.15), so as to obtain the permanent displacements in different dynamic cases.

Table 6.5 Factor of safety in static and pseudo-static cases

Case 1		Case 2	Case 3 (upward-vertical)	Case 3 (downward-vertical)
1.16	0.3PGA	0.76	0.79	0.76
	0.5PGA	0.65	0.66	0.65

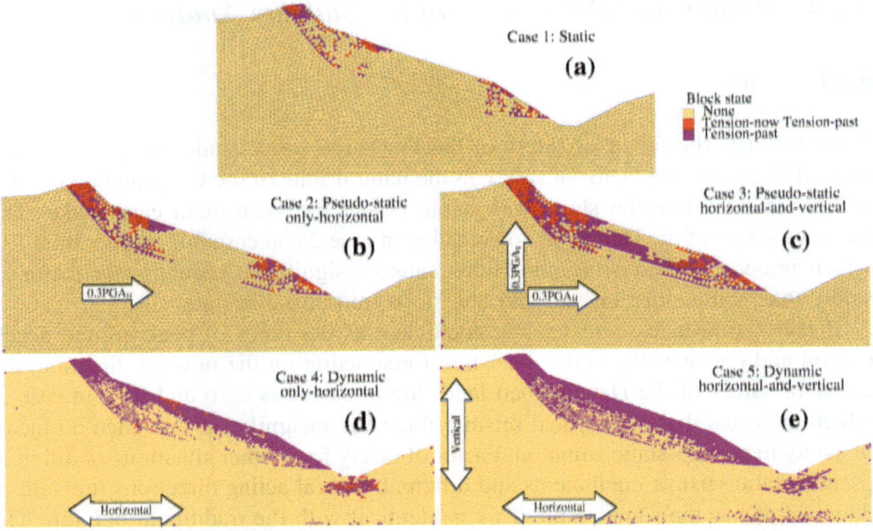

Fig. 6.19 Distributions of tension failure area in case 1, 2, 3, 4, and 5

Firstly, validation of the accuracy of dynamic analysis is carried out. Output displacement time histories of stable monitoring point 1# in case 4 and case 5 (Fig. 6.20c, d) are compared to the displacement histories integrated from the input acceleration records (Fig. 6.20a, b). For case 4, from Fig. 6.20c, d, we can find that the horizontal displacement is consistent well with the input loading, while the vertical displacement has a small value which is different from the input one since

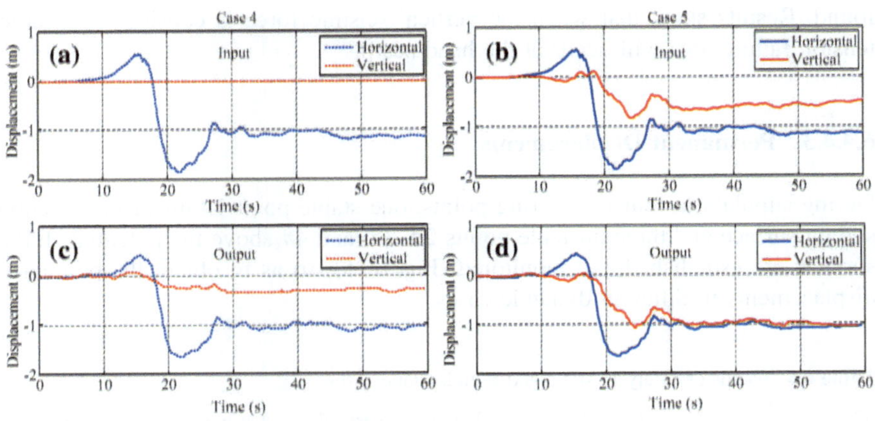

Fig. 6.20 Validation of the permanent displacement. Displacement time histories integrated from input acceleration record in **a** case 4 and **b** case 5; permanent displacement time histories of monitoring point (1#) under the sliding surface in **c** case 4 and **d** case 5

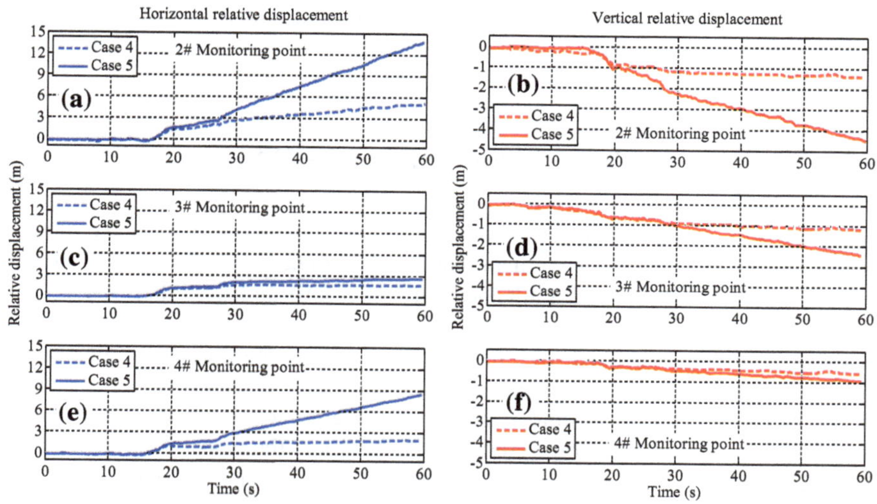

Fig. 6.21 Relative displacement of monitoring points (2#, 3#, and 4#) above the sliding surface in case 4 and case 5

the reflection and composition of the seismic wave. For case 5, from Fig. 6.19b, d, we can find that both the horizontal and vertical displacements are consistent well with the input ones except a little difference in vertical displacement with the same reason as case 4. In addition, the displacement time histories are convergent which confirms the stability of the monitoring point under the sliding surface.

Then, horizontal and vertical relative displacements of monitoring points above the sliding surface in case 4 and case 5 are presented. In here, the relative displacement indicates the difference value between the study point and the stable monitoring point 1#, i.e., $D_{\text{relative}} = D_{\text{study}} - D_{\text{stable}}$. From Fig. 6.21, it can be found that both the horizontal and vertical relative displacements of all monitoring points above the sliding surface are very different between case 4 and case 5. The relative displacements of three monitoring points in case 5 are larger than those in case 4. In addition, all the relative displacement histories in Fig. 6.21 are non-convergent which indicates the slope is instable under the earthquake loading with or without the vertical seismic force.

Meanwhile, Table 6.6 lists the relative displacements of monitoring points in both horizontal and vertical directions. From Table 6.6, it also can be found that the

Table 6.6 Relative displacements of monitoring points in case 4 and case 5 (*unit* m)

	1#		2#		3#		4#	
	Case 4	Case 5	Case 4	Case 5	Case 4	Case 5	Case 4	Case 5
Horizontal	−1.02	−1.03	5.28	14.11	1.91	2.88	2.07	8.91
Vertical	−0.26	−1.02	−1.13	−4.33	−0.55	−2.10	−0.45	−0.82

permanent displacements in case 5 are much larger than those in case 4 except the stable point 1#.

6.5 Run-Out Analysis by the Extended DDA

Seismic DDA can successfully simulate the movement of earthquake-induced landslide (Wu et al. 2009). Two main features determine the Daguangbao landslide is a unique case, one is near-fault location ($\approx$6.5 km) and the other one is huge scale ($\approx 800 \times 10^6$ m^3). The near-fault location determines the Daguangbao landslide must be shocked by the extreme ground motion from the strong Wenchuan earthquake. And the Daguangbao landslide located on the meizoseismal area where the vertical seismic component is very large (surface displacement shown in Fig. 6.9 provides some credence). In addition, the landslide is so huge that the size effects must be considered. The friction coefficient measured in the laboratory is no longer suitable for stability analysis and run-out analysis.

To these two features, the Daguangbao landslide is simulated by the newest seismic DDA code in which multi-direction seismic forces can be applied in the base block directly, and experience equations based on Hoek–Brown failure criterion are applied to back-calculate the material strength by trying to consider the size effect.

6.5.1 *The Seismic DDA Code and Basic Assumptions*

Both DEM and DDA employ the equations of dynamic motion which are solved at finite points in time, in a series of time steps, but there are similarities in their formulations but significant differences of the solution schemes and contact mechanics. In the solution schemes, equations of motion in DDA are derived using the principle of minimization of the total potential energy of the system, while the equations of motion as implemented in DEM are derived directly from the force balance equations, which still resultant unbalanced force after a time step and damping is necessarily used to dissipate energy. In the contact mechanics, the DDA used a penalty method in which the contact is assumed to be rigid. No overlapping or interpenetration of the blocks is allowed as same as real physical cases, whereas soft contact approach is used in DEM. The soft contact approach requires laboratory or field measured joint stiffness, which may be difficult to obtain in many cases. Many comparisons of basic models (sliding, colliding, and rolling models) between the DEM and DDA were carried out and show that the results from DDA are more close to the analytical values than that from DEM (Zheng 2010).

DDA is a dynamic numerical analysis method capable of evaluating the impacts area of an earthquake-induced landslide when seismic impacts are integrated into simulations. Hatzor and Feintuch (2001) are the first to validate the use of DDA in simulating dynamic landslides by studying the dynamic of block sliding on an inclined plane, in which they assume that the base block is fixed and earthquake accelerations are directly considered as body force and added to the sliding block in DDA. Based on the same input model of seismic loadings, Makris and Roussos (2000), Shi (2002), Kong and Liu (2002), Ishikawa et al. (2002), Hatzor et al. (2004), Tsesarsky et al. (2005), Yagoda and Hatzor (2010), Bakun-Mazor et al. (2012) studied the dynamic response or/and stability analysis of tunnel, slope, dam, foundation, or ancient masonry structure using DDA. Alternatively, Sasaki et al. (2004) developed an acceleration input method different from the original DDA algorithm to simulate the dynamic behavior of a slope with sliding block. In his method, the seismic accelerations were applied to the base block, which is different from the former seismic loadings input model. Sasaki et al. (2007) applied the same earthquake input model to analyze several cases of simple block structures under harmonic accelerations to acquire the relationships between natural frequencies of elastic block structures and applied accelerations. Later, Wu group (2009, 2010, 2011, 2011) applied the DDA to simulate the kinematic behavior of sliding rock blocks in the Tsaoling landslide and the Chiu-fen-erh-shan landslide induced by the 1999 Chi-Chi earthquake. In this paper, we apply the latest DDA code, in which multi-direction seismic forces can be considered, to study the effects of vertical seismic force on run-out of earthquake-induced landslide.

DDA can consider the seismic loadings by three forms: acceleration, velocity, and displacement time histories. Loading of time-dependent acceleration is widely used by many researchers to conduct earthquake analysis (e.g., Wu 2010; Hatzor and Feintuch 2001; Sasaki et al. 2004; Ning and Zhao 2012). In practice, acceleration loadings could be used to the failure mass (method 1) or the base (method 2) as volume forces. For simulation of earthquake-induced landslide, there are three main motion types for failure mass: sliding, rolling, and collision. For sliding motion type, Wu (2010) has verified that incorrect absolute movements of sliding and based blocks will be obtained if the acceleration time history is applied to failure mass (method 1). In addition, collision motion means the failure block may separate from the base. In this situation, it is senseless if the seismic loadings are still applied to the failure mass. Although sliding is the main motion type for a landslide, collision is often appeared due to violent collapse or microtopography of slide bed. Comparatively, there will be no such problem when the method 2 is used. Hence, method 2 was used here to simulate the Daguangbao landslide.

In addition, when the acceleration time history is applied to the base, the base may get a residual velocity at the end of the seismic loadings, although the ground vibration should be stopped, thus a continuing displacement shift will possibly appear. Wu (2010) presented an approach to solve this problem by setting the velocity of the base as zero fictitiously when the seismic loadings end. The seismic subroutine combined the zero-residual-velocity approach is programmed in the latest version of the DDA code from Gen-Hua Shi and used in this study.

When applying acceleration time history to the base, the vertical constraint can be set to avoid up–down movement if only the horizontal seismic loading is considered. Once the vertical seismic force is also applied to base, no constraint point can be set to it. Thus, the base must be assumed to be not submitted to gravity, and the density of the base must assumed to be significantly large as well (Sasaki et al. 2004), so that the accelerations of the base driven by the contact forces between the sliding mass and the base can be neglected. In physics, we can regard the base block as concentration of the whole earth, and the gravity is the internal force that cannot be considered in force analysis. Although a large-scale landslide can induce certain ground motion (called landquake by Favreau et al. 2010), the induced motion is very small that can be neglected, e.g., the order of velocity induced by a 2.5×10^6 m^3 landslide is 10^{-6} m/s (Favreau et al. 2010).

Because of the restriction of numerical simulation technology, only a 2-D simulation was carried out. Certain assumptions were made to capture the main features of the large rock avalanche.

1. Blocks are treated as 2-D polygons;
2. Sliding masses are assumed to be divided to smaller blocks by preexisting joints;
3. Geometry of the slope model is simplified;
4. Earthquake forces only act to the base block.
5. Density of base block is assumed as 10^5 times to avoid the influences of gravity of sliding blocks
6. Gravity acceleration acting to the base block is set to zero to make up–down free vibrations.

Assumptions (1)–(4) are often used in numerical simulation and widely accepted. Specially, in order to check the rationality of the assumptions (5) and (6), three cases for a sliding block model (Fig. 6.22a) are carried out and an idealized one-cycle sinus pulse (Fig. 6.22b) is used as excitation:

Case 1 the base block has a virtual density (10^5 times of its real value, $M = 10^5\ m$) and not submitted to gravity;
Case 2 the base block has a realistic density ($M = m$) and not submitted to gravity;
Case 3 the base block has a realistic density ($M = m$) and submitted to gravity.

The theory solution for the dynamic behavior (acceleration, velocity, and displacement) of studied model can be obtained from the BLOCK program presented in Chap. 4 (Zhang et al. 2013a). Comparisons of relative velocity and displacement between theory solution and different DDA results are shown in Fig. 6.23. From Fig. 6.23, we can find that Case 1 gives convergent results comparing to Case 2. The not convergent results from Case 2 are just induced by the non-ignorable contact forces between the sliding block and the base because the same density is set to both sliding block and base. It should be noted that the DDA results of Case 1 are little smaller than the theory solutions. That is because the rigid-plastic assumption is used in the derivation of theory solutions, whereas elastic–plastic model is used in the DDA program. Case 3 gives a complete different results comparing to other results, that is because the gravity of the whole sliding system is out of use, i.e., the derive forces of sliding block is disappeared. Hence, the larger

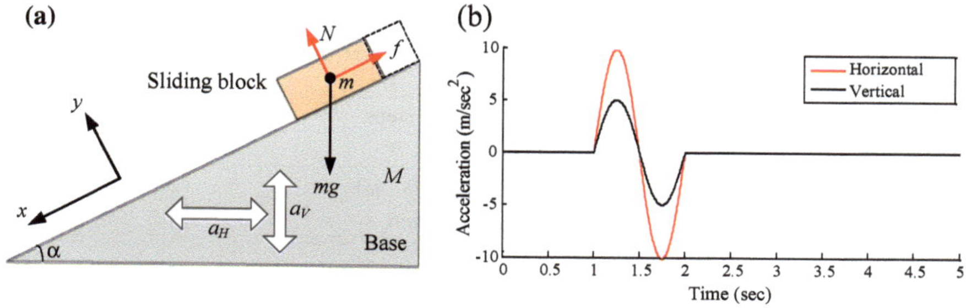

Fig. 6.22 **a** Dynamic sliding block model; and **b** excitation

density and zero gravity of base have no influence on the simulation results and give the best results comparing to other cases.

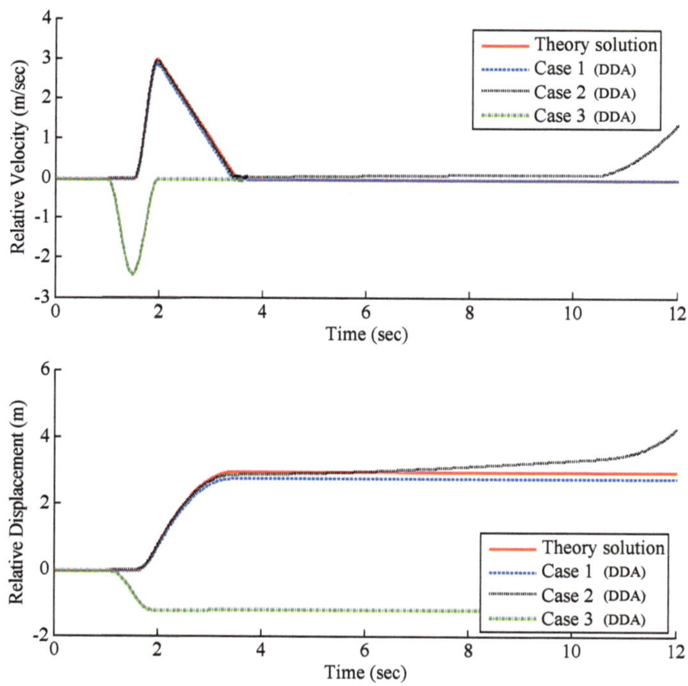

Fig. 6.23 Comparison of velocity and displacement of sliding block between theory solution and different DDA results

Table 6.7 Control parameters for DDA

Parameter	Value
Assumed maximum displacement ratio (g_2)	0.001
Total number of time steps	20,000
Time step (g_1)	0.005 s
Contact spring stiffness (g_0)	5.0×10^8 kN/m
Factor of over-relaxation	1.3

6.5.2 Geometry and Properties of Sliding Blocks

The main sliding direction of the Daguangbao landslide, N60°E, is selected as analysis profile. The DDA model is depicted in Fig. 6.24. In this simulation, based on the shape of failure surface and the character of slope topography, the whole slope is divided into three parts: base block, upper sliding mass, and lower sliding mass. Then, two sliding masses are divided into the smaller discrete deformable blocks based on preexisting discontinuities. Strength parameters, analytical conditions and material properties are identical to those used in the stability analysis. The control parameters used in DDA are shown in Table 6.7. The slope system is free at both horizontal and vertical directions.

6.5.3 Earthquake Forces

Figure 6.25 shows the input combined acceleration records. Velocity and displacement time histories are obtained by once and twice integration from acceleration record, respectively. The residual values of displacement time histories indicate that the Daguangbao slope moved in that direction during the earthquake. The horizontal earthquake wave is the projection combination in the main sliding

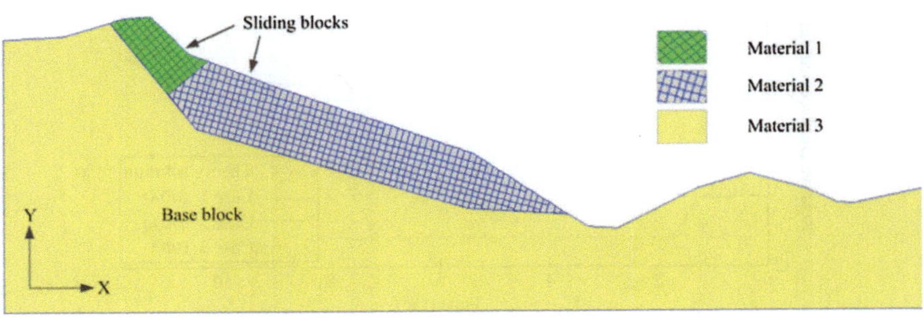

Fig. 6.24 DDA model of the Daguangbao landslide

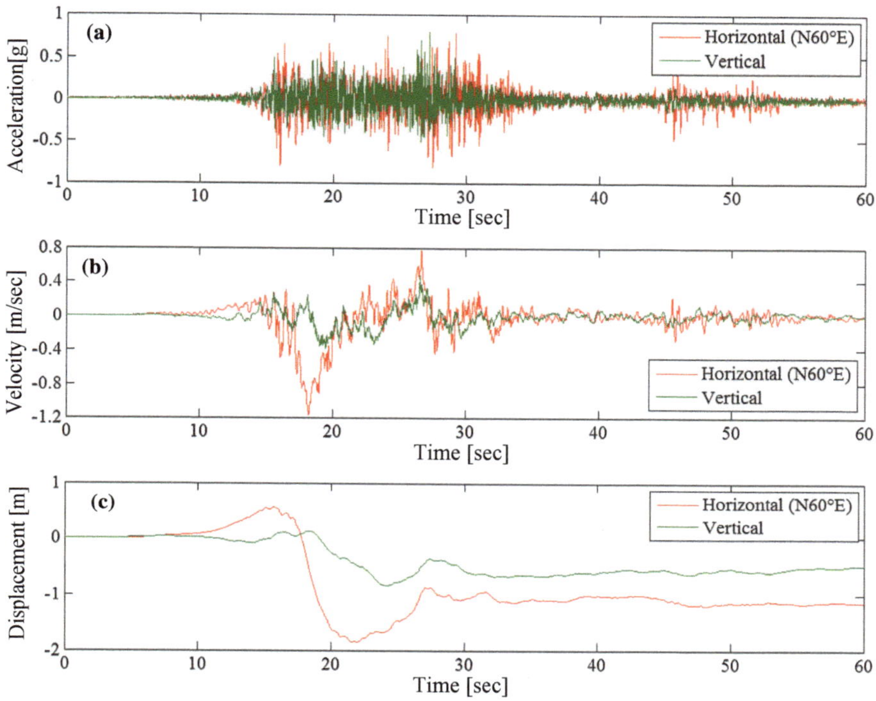

Fig. 6.25 **a** Input post-corrected horizontal and vertical ground acceleration records, **b** velocity, and **c** displacement histories projected to N60°E direction

direction (N60°E) using the MZQP acceleration records in E–W and N–S directions as Eq. 6.12. The inputted vertical earthquake wave is the MZQP acceleration records in U–D direction. Because the seismic data of the first 20 s shown in Fig. 6.9 are pre-seismic, the duration of earthquake wave is 60 s (20–80 s in Fig. 6.9). The horizontal and vertical acceleration records are acted to the base block at the same time.

6.5.4 Results and Discussions of the Run-out Analysis

6.5.4.1 Simulation Results

Figure 6.26 shows the horizontal and vertical dynamic behavior (velocity and displacement time histories) of the base while both horizontal and vertical seismic forces are applied to the base. Clearly, the calculated time step exceeds the duration of seismic data. In the duration of the seismic loadings (0–60 s), the computational results coincide well with input seismic data. In fact, there is a nonzero velocity in

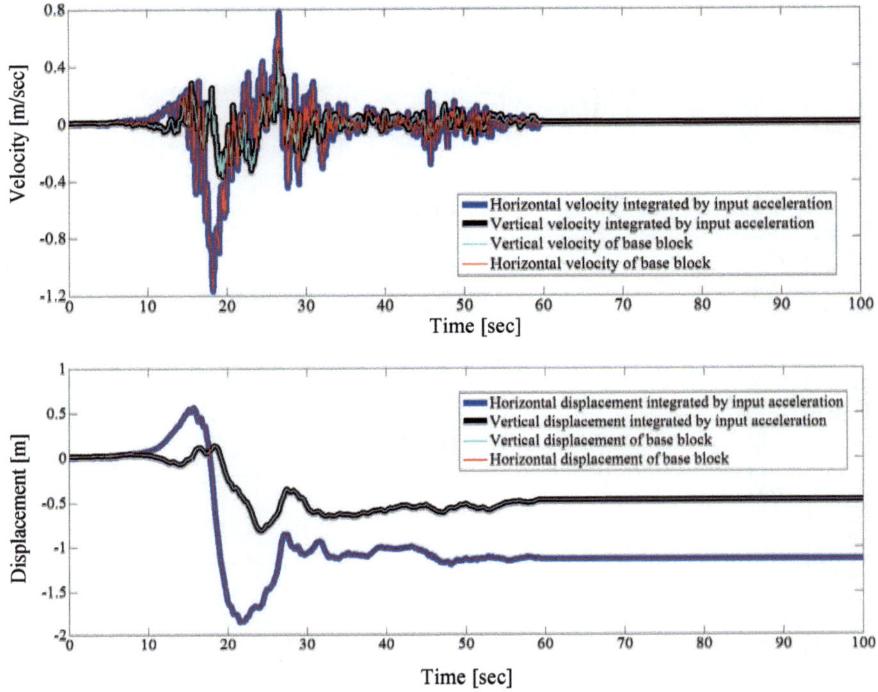

Fig. 6.26 Horizontal and vertical velocities and displacements of base block

the end of inputted seismic data, which may be carried forward by the base in the DDA simulation. Thus, a displacement shift with a gradient of the nonzero velocity will be appeared. So, the zero-residual-velocity approach stated in previous section is used to set the velocity of the base at zero after the end time of the seismic data. Thus, the velocity and displacement time histories of the base are horizontal lines.

Figure 6.27 shows the post-failure behavior of the Daguangbao landslide simulated by the seismic DDA code. Simulated results show that the sliding blocks climb over the Pingliangzi. After overlapping the final step of DDA calculation with the topographic cross section at the Daguangbao landslide (Fig. 6.6), the deposit pattern of the simulated Daguangbao landslide under horizontal-and-vertical situation coincides well with local topography.

6.5.4.2 Role of the Seismic Loadings on Landslide Run-Out

The estimation of the movement behaviors of a potential landslide is very important to mitigate the landslide disaster. Especially, the run-out distance is one of the major parameters in landslide risk assessment and preventive measure design. Long run-out is one of the major characteristics of earthquake-induced landslides. Many

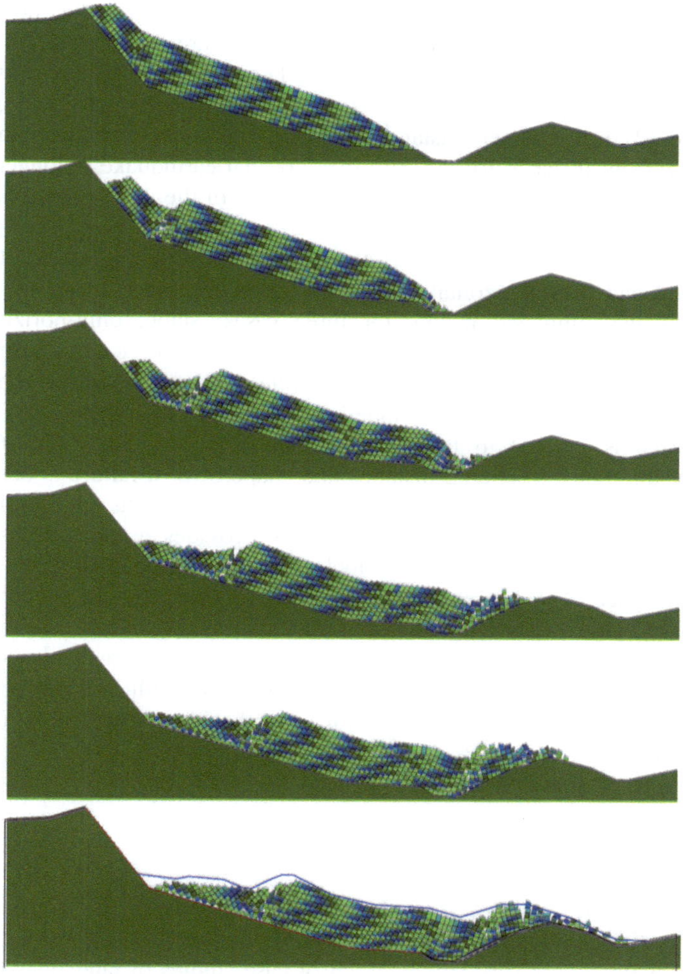

Fig. 6.27 Simulation results of the Daguangbao landslide

researchers made great efforts to understand how and why large falling masses of rock can move unusually long run-out distance. Researchers repeatedly revisited the problem using a wide variety of approaches. These efforts yielded more than 20 mechanical models for explaining long run-out in high-volume rapid landslides. Actually, most of the existed models are helpful in the estimation of the run-out distance. However, very less of them considered the earthquake dynamical behavior. For this reason, we proposed a model called multiplex acceleration model (MAM) to explain the long run-out mechanism, and model tests using shaking table were carried out to investigate the effects of earthquake on the movement of landslide (Chen et al. 2011). The main conclusions are as follows: (1) Effect of earthquake on the movement distance is very large, as the movement distance for the case of a 400 gal earthquake is about 3 times longer than the case of no earthquake; (2) the movement distance is proportional to the earthquake magnitude, as the movement distance for the case of a 400 gal earthquake is longer than the case of a 200 gal earthquake; (3) the size and shape of the falling block have effect on the movement distance.

In order to validate the effects of earthquake on landslide run-out, additional two cases, static and only-horizontal, were carried out. Figure 6.28 shows the comparisons of the simulated results for three cases: static, only-horizontal, and horizontal-and-vertical. In static situation (Fig. 6.28a), the slope is overall stable except a little settlement in the top of slope (circle zone), which consists with that of limit equilibrium method, i.e., the slope is stable before the earthquake. When the seismic force is applied to the base, the slope is not stable (Fig. 6.28b, c). Significant difference in sliding distance and shape of post-failure is observed for the two different dynamic situations. The simulation results of only-horizontal situation present that the Huangdongzi valley provides a moat to prevent the overtopping of sliding rocks. The toe block of the Daguangbao landslide generates a sliding distance of 944 m at horizontal direction. Whereas, simulated results of the horizontal-and-vertical situation show that the sliding blocks climb over the Pingliangzi and the travel distance of toe block is 1901 m. These results confirm the existed conclusions that the earthquake has a significant influence on the landslide initiation and run-out. In addition, effect of vertical seismic force on the movement distance is large.

6.5.4.3 Sensitivity Analysis of Shear Strength

As the Coulomb friction law is used in most numerical models, both continuous model and discontinuous model, the sensitivity of landslide deposit to friction coefficient is obviously very strong. A lot of case studies of landslides have been done using continuous models, e.g., SHALTOP (e.g., Favreau et al. 2010; Kuo et al. 2009), DAN 3D (e.g., McDougall and Hungr 2004; Sosio et al. 2012; McDougall 2006), MADFlow and its extensions (e.g., Chen et al. 2006; Crosta and Agliardi 2003; Crosta and Frattini 2003; Crosta et al. 2004, 2005, 2007, 2009a, b), and discontinuous models, e.g., DEM (e.g., Latha and Garaga 2010), DDA (e.g., Wu et al. 2005; Wu and Chen 2011; Wu 2007). All of these works confirm the strong influence of friction coefficients on the landslide run-out.

The friction coefficients used in this simulation are back-calculated from experience equations of Hoek–Brown. Some associated uncertainties are contained, so the shear parameters used in Table 6.3 are not so precise. Hence, a sensitivity

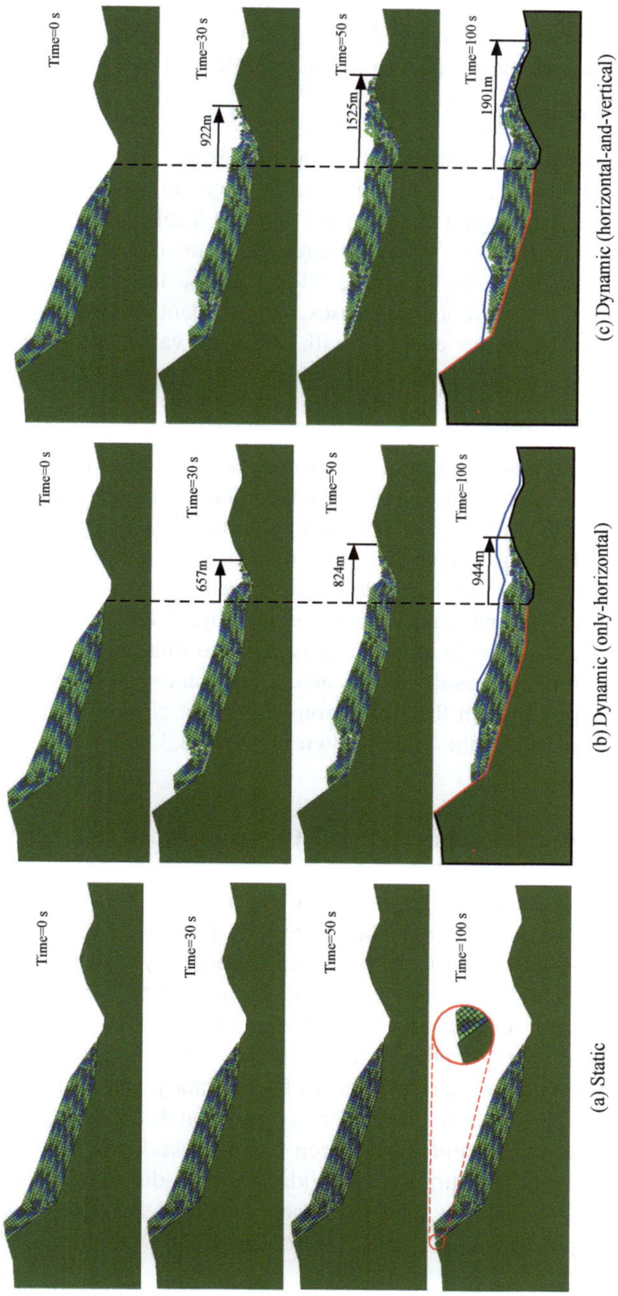

Fig. 6.28 Post-failure behavior of the Daguangbao landslide for three cases

analysis of them, c and φ, is carried out here. The cohesion and friction angle used in sensitivity analysis, c' and φ', are determined by two strength factors K_c and K_φ as $c' = K_c\, c$ and $\tan\varphi' = K_\varphi \tan\varphi$. 0.8, 1.0, 1.3, and 1.5 of K_c and K_φ are used to calculate the cohesion and friction angle, respectively.

(a) **Role of cohesion**

Four cases for different cohesions calculated from four strength factors are carried out to study the effects of cohesion on the slope failure process. In this series of analysis, the friction angle is keep as the value in Table 6.3. Figure 6.29 shows the results of the simulations. When the strength factor is 1.5, the slope can keep stable (Fig. 6.29d). With the decrease of cohesion, the landslide is initiated (Fig. 6.29a–c). From the front three cases, insignificant difference in the failure process is observed for the analysis with different values of cohesion, which illustrates the effects of cohesion on the landslide run-out are little.

(b) **Role of friction angle**

Four cases for different friction angles were used to study the influences of the friction angle on the landslide run-out, while the cohesion is maintained as shown in Table 6.3. Figure 6.30 shows the results of the simulations. When the strength factor is 1.5, the slope can keep stable (Fig. 6.30d). With the decrease of cohesion, the landslide is initiated (Fig. 6.30a–c). A significant change of slope failure process can be found. The simulated deposits on the zone 1 and zone 2 topographical indicate that the friction angle strongly governed the final deposition topography. Comparing the simulated results to the actual post-landslide topography, we can find that the friction angle in the simulations that most closely matched the actual depositional topography is the value shown in Table 6.3.

6.5.4.4 Effects of Topography and 3-D Effects of Block Geometry

It should be noted that the topography has strong influence on the landslide run-out and deposit, and it was studied in the literature (e.g., Favreau et al. 2010; Zhang et al. 2013b, c). In addition, the 3-D effects of block geometry also play a key role in landslides deposit and dynamic (Phillips et al. 2006; Chen et al. 2011). A three-dimensional method should be used to take these effects into account. Some widely used thin layer depth-averaged models were developed into three-dimensional versions, in which the effects of topography and 3-D effects of block geometry can be considered. However, conventional continuum approaching models, which neglect the contacts between rocks, make it impossible to trace the position of individual rock during a landslide. The two-dimensional DDA method used here cannot consider the effect of topography and the 3-D effects of block geometry. To this point, an impeccable three-dimensional DDA should be developed in future.

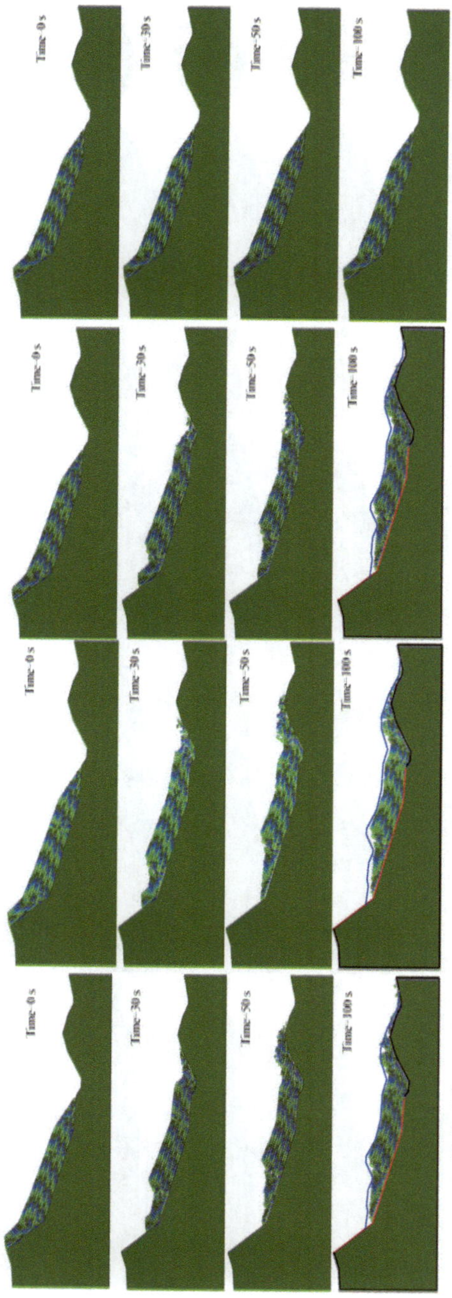

Fig. 6.29 Effects of cohesion on landslide run-out

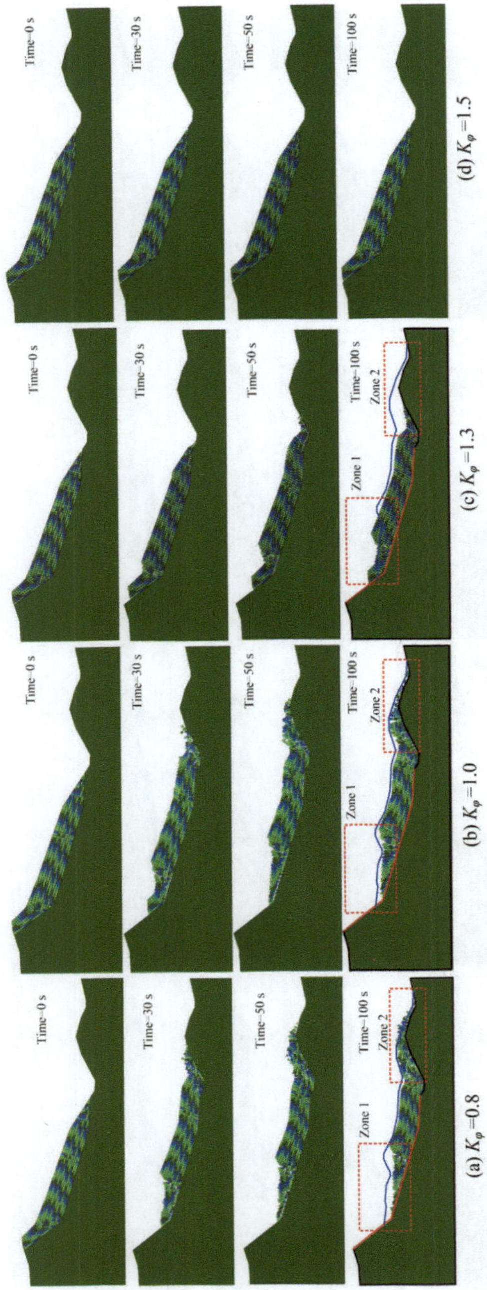

Fig. 6.30 Effects of friction angle on landslide run-out

6.6 Conclusions

With the aim of investigating the effect of vertical earthquake force on the initiation of the earthquake-induced landslide, a case of the Daguangbao landslide induced by the Wenchuan earthquake is presented in this paper. Firstly, two key issues of seismic stability analysis of near-fault large-scale landslide are considered. One is how to select and correct the input near-fault ground motion, and the other is how to estimate the material strength considering the effect of slope scale. For the former, three selection criteria and multi-linear baseline correction are proposed, and for the later, nonlinear Hoek–Brown failure criterion is used to estimate the material strength in which the slope scale can be considered. Secondly, five cases: (1) static, (2) pseudo-static only-horizontal, (3) pseudo-static horizontal-and-vertical, (4) dynamic only-horizontal, and (5) dynamic horizontal-and-vertical are performed using finite difference program FLAC^{3D}, under the real seismic waves near the study site. The results show that the seismic conditions cause a significant reduction in factor of safety than static situation. It also found that the vertical seismic has a significant influence on tension failure of block, although it has an insignificant influence on change of the factor of safety. Another important conclusion is the effect of vertical seismic force on relative displacement of potential sliding mass is significant. In addition, large area of tension failure caused by the combined seismic forces at back edge of the slope applies the evidence of effect of vertical seismic force on failure mechanism of slope stability.

In traditional approaches, only horizontal earthquake force is considered while the vertical force is neglected. The results of this paper show that it is necessary to take the effect of vertical seismic force into consideration, especially when the study site is located in meizoseismal areas.

Post-failure discrete analysis of the Daguangbao landslide is carried out using latest seismic DDA code. In order to get the same response between input and output vertical acceleration of earthquake record, we assumed the density of base block is much larger than its real value and the base block out of the gravity. In this study, in order to consider the size effect of the run-out analysis, Hoek–Brown experience equations, which can consider the influence of slope height on equivalent shear strength, are used for back-calculate the shear strength. In spite of some assumptions, the study is able to reproduce a post-failure configuration resembling that observed in the field.

A comparison of simulation results from three situations, static, only-horizontal, and horizontal-and-vertical, is carried out. Seismic force has a significant influence on the arrival distance and shape of post-failure. Arrival distance from horizontal-and-vertical situation is larger than that from only-horizontal situation. In addition, the deposit pattern of the simulated Daguangbao landslide under horizontal-and-vertical situation coincides well with local topography. The vertical seismic force should be considered for landslide assessment and management, especially in the situation that the studied site located on the meizoseismal area during the earthquake.

A series of sensitivity analyses are carried out to study the effects of cohesion and friction angle on landslide run-out. The simulation results illustrate that the effect of cohesion on the landslide run-out is little, while the friction angle has a significant influence on the landslide run-out.

References

Bakun-Mazor, D., Hatzor, Y. H., & Glaser, S. D. (2012). Dynamic sliding of tetrahedral wedge: The role of interface friction. *International Journal for Numerical and Analytical Methods in Geomechanics, 36*(3), 327–343.

Boore, D. M. (2001). Effect of baseline corrections on displacements and response spectra for several recordings of the 1999 Chi-Chi, Taiwan, earthquake. *Bulletin of the Seismological Society of America, 91*(5), 1199–1211.

Chang, K.-J., Taboada, A., Lin, M.-L., & Chen, R.-F. (2005). Analysis of landsliding by earthquake shaking using a block-on-slope thermo-mechanical model: Example of Jiufengershan landslide, central Taiwan. *Engineering Geology, 80*(1–2), 151–163.

Chen, H., & Lee, C. (2000). Numerical simulation of debris flows. *Canadian Geotechnical Journal, 37*(1), 146–160.

Chen, Z. Y., Yin, J. H., & Wang, Y. J. (2006). The three-dimensional slope stability analysis recent advances and a forward look. In Advances in Earth Structures: Research to Practice.

Chen, G., Li, Y., Zhang, Y., & Wu, J. (2011). Earthquake induced a chain disasters. In Earthquake Research and analysis—Statistical studies, Observations and Planning. InTech, pp. 383–416.

Crosta, G., & Agliardi, F. (2003). A methodology for physically based rockfall hazard assessment. *Natural Hazards and Earth System Science, 3*(5), 407–422.

Crosta, G., & Frattini, P. (2003). Distributed modelling of shallow landslides triggered by intense rainfall. *Natural Hazards and Earth System Science, 3*(1/2), 81–93.

Crosta, G. B., Chen, H., & Lee, C. F. (2004). Replay of the 1987 Val Pola landslide, Italian Alps. *Geomorphology, 60*(1–2), 127–146.

Crosta, G. B., Frattini, P., & Fusi, N. (2007). Fragmentation in the Val Pola rock avalanche, Italian Alps. Journal of Geophysical Research: Earth Surface (2003–2012), 112(F1).

Crosta, G. B., Imposimato, S., & Roddeman, D. (2009a). Numerical modelling of entrainment/deposition in rock and debris-avalanches. Engineering Geology, *109*(1–2), 135–145.

Crosta, G. B., Imposimato, S., & Roddeman, D. (2009b). Numerical modeling of 2-D granular step collapse on erodible and nonerodible surface. Journal of Geophysical Research, *114*(F3).

Crosta, G., Imposimato, S., Roddeman, D., Chiesa, S., & Moia, F. (2005). Small fast-moving flow-like landslides in volcanic deposits: the 2001 Las Colinas Landslide (El Salvador). *Engineering geology, 79*(3), 185–214.

Cui, P., Chen, X.-Q., Zhu, Y.-Y., Su, F.-H., Wei, F.-Q., Han, Y.-S., et al. (2009). The Wenchuan Earthquake (May 12, 2008), Sichuan Province, China, and resulting geohazards. *Natural Hazards, 56*(1), 19–36.

Cundall, P. (1971). A computer model for simulating progressive, large scale movements in blocky rock system. *In Symposium of International Society of Rock Mechanics* (pp. 11–18). France: Nancy.

Davies, T. R., & McSaveney, M. J. (2009). The role of rock fragmentation in the motion of large landslides. *Engineering Geology, 109*(1–2), 67–79.

Dawson, E. M., Roth, W. H., & Drescher, A. (1999). Slope stability analysis by strength reduction. *Géotechnique, 49*(6), 835–840.

de Michele, M., Raucoules, D., de Sigoyer, J., Pubellier, M., & Chamot-Rooke, N. (2010). Three-dimensional surface displacement of the 2008 May 12 Sichuan earthquake (China) derived from Synthetic Aperture Radar: Evidence for rupture on a blind thrust. *Geophysical Journal International, 183*(3), 1097–1103.

Denlinger, R. P., & Iverson, R. M. (2001). Flow of variably fluidized granular masses across three dimensional terrain: 2. Numerical predictions and experimental tests. Journal of Geophysical Research: Solid Earth (1978–2012), *106*(B1), 553–566.

Elgamal, A.-W. M., Scott, R. F., Succarieh, M. F., & Yan, L. (1990). La Villita dam response during five earthquakes including permanent deformation. *Journal of Geotechnical Engineering, 116*(10), 1443–1462.

Erismann, T. H., & Abele, G. (2001). Dynamics of rockslides and rockfalls. Springer.

Favreau, P., Mangeney, A., Lucas, A., Crosta, G., & Bouchut, F. (2010). Numerical modeling of landquakes. Geophysical Research Letters, *37*(15).

Gorum, T., Fan, X., van Westen, C. J., Huang, R. Q., Xu, Q., Tang, C., et al. (2011). Distribution pattern of earthquake-induced landslides triggered by the 12 May 2008 Wenchuan earthquake. *Geomorphology, 133*(3–4), 152–167.

Griffiths, D., & Prevost, J. H. (1988). Two-and three-dimensional dynamic finite element analyses of the Long Valley Dam. *Geotechnique, 38*(3), 367–388.

Hatzor, Y. H., & Feintuch, A. (2001). The validity of dynamic block displacement prediction using DDA. International Journal of Rock Mechanics & Mining Sciences.

Hatzor, Y., Arzi, A. A., Zaslavsky, Y., & Shapira, A. (2004). Dynamic stability analysis of jointed rock slopes using the DDA method: King Herod's Palace, Masada, Israel. *Int J Rock Mech Min Sci, 41*(5), 813–832.

Havenith, H. B., & Bourdeau, C. (2010). Earthquake-induced landslide hazards in mountain regions. A review of case histories from central asia. *Geologica Belgica, 13*(3), 137–152.

Hoek, E. (1990). Estimating Mohr-Coulomb friction and cohesion values from the Hoek-Brown failure criterion. *International Journal of Rock Mechanics and Mining Sciences, 27*(3), 227–229.

Hoek, E., & Brown, E. T. (1988). The Hoek-Brown failure criterion-a 1988 update. In 15th Canada Rock Mechanics Symposium, pp. 31–38.

Hoek, E., & Brown, E. T. (1997). Practical estimates of rock mass strength. *International Journal of Rock Mechanics and Mining Sciences, 34*(8), 1165–1186.

Hoek, E., Carranza-Torres, C., & Corkum, B. (2002). Hoek-Brown failure criterion—2002 edition. In NARMS-TAC Conference, Toronto, pp. 267–273.

Hsü, K. J. (1975). Catastrophic debris streams (sturzstroms) generated by rockfalls. *Geological Society of America Bulletin, 86*(1), 129–140.

Huang, R., Pei, X., & Li, T. (2008). Basic characteristics and formation mechanism of the largest scale landslide at Daguangbao occurred during the Wenchuan earthquake. *Journal of Engineering Geology, 16*(6), 730–741.

Huang, R., Xu, Q., & Huo, J. (2011). Mechanism and geo-mechanics models of landslides triggered by 5.12 Wenchuan Earthquake. *Journal of Mountain Science, 8*(2), 200–210.

Huang, R., Pei, X., Zhang, W., Li, S., & Li, B. (2009). Further examination on characteristics and formation mechanism of Daguangbao landslide. *Journal of Engineering Geology, 17*(6), 725–736.

Huang, R., Pei, X., Fan, X., Zhang, W., Li, S., & Li, B. (2012). The characteristics and failure mechanism of the largest landslide triggered by the Wenchuan earthquake, May 12, 2008, China. *Landslides, 9*(1), 131–142.

Ishikawa, T., Sekine, E., & Ohnishi, Y. (2002). Shaking table tests of coarse granular materials with discontinuous analysis. Proc. of ICADD-5, BALKEMA, pp. 181–187.

Itasca. (2007). Fast Lagrangian Analysis of Continua in 3-Dimension (FLAC3D V3.1). Itasca Consulting Group, Minnesota.

Iwan, W. D., Moser, M. A., & Peng, C.-Y. (1985). Some observations on strong-motion earthquake measurement using a digital accelerograph. *Bulletin of the Seismological Society of America, 75*(5), 1225–1246.

Keefer, D. K. (2002). Investigating landslides caused by earthquakes—A historical review. *Surveys in geophysics, 23,* 473–510.

Kong, X., & Liu, J. (2002). Dynamic failure numeric simulations of model concrete-faced rock-fill dam. *Soil Dynamics and Earthquake Engineering, 22*(9–12), 1131–1134.

Kramer, S. L. (1996). Geotechnical earthquake engineering. Prentice-Hall Civil Engineering and Engineering Mechanics Series (p. 1). Upper Saddle River, NJ: Prentice Hall.

Kuo, C. Y., Tai, Y. C., Bouchut, F., Mangeney, A., Pelanti, M., Chen, R. F., et al. (2009). Simulation of Tsaoling landslide, Taiwan, based on Saint Venant equations over general topography. *Engineering Geology, 104*(3–4), 181–189.

Latha, G. M., & Garaga, A. (2010). Seismic stability analysis of a Himalayan rock slope. *Rock Mechanics and Rock Engineering, 43*(6), 831–843.

Lee, K. L. (1974). *Seismic permanent deformation in earth dams*. Los Angeles, CA: University of California.

Makris, N., & Roussos, Y. (2000). Rocking response of rigid blocks under near-source ground motions. *Geotechnique, 50*(3), 243–262.

McDougall, S. (2006). A new continuum dynamic model for the analysis of extremely rapid landslide motion across complex 3D terrain, University of British Columbia.

McDougall, S., & Hungr, O. (2004). A model for the analysis of rapid landslide motion across three-dimensional terrain. *Canadian Geotechnical Journal, 41*(6), 1084–1097.

Moretti, L., Mangeney, A., Capdeville, Y., Stutzmann, E., Huggel, C., Schneider, D., et al. (2012). Numerical modeling of the Mount Steller landslide flow history and of the generated long period seismic waves. Geophysical Research Letters, *39*(16).

Ning, Y., & Zhao, Z. (2012). A detailed investigation of block dynamic sliding by the discontinuous deformation analysis. International Journal for Numerical and Analytical Methods in Geomechanics, 1–21.

Phillips, J., Hogg, A., Kerswell, R., & Thomas, N. (2006). Enhanced mobility of granular mixtures of fine and coarse particles. *Earth and Planetary Science Letters, 246*(3–4), 466–480.

Prevost, J. H. (1981). DYNA-FLOW: A nonlinear transient finite element analysis program. Princeton University, Department of Civil Engineering, School of Engineering and Applied Science.

Sasaki, T., Hagiwara, I., Sasaki, K., Ohnishi, Y., & Ito, H. (2007). Fundamental studies for dynamic response of simple block structures by DDA. *In Proceedings of the eighth international conference on analysis of discontinuous deformation: fundamentals and applications to mining & civil engineering* (pp. 141–146). China: Beijing.

Sasaki, T., Hagiwara, I., Sasaki, K., Yoshinaka, R., Ohnishi, Y., & Nishiyama, S. (2004). Earthquake response analysis of rock-fall models by discontinuous deformation analysis. *In Proceedings of the ISRM international symposium 3rd ARMS* (pp. 1267–1272). Japan: Kyoto.

Savage, S., & Hutter, K. (1989). The motion of a finite mass of granular material down a rough incline. *Journal of fluid mechanics, 199*(1), 177–215.

Scheidegger, A. E. (1973). On the prediction of the reach and velocity of catastrophic landslides. *Rock Mechanics, 5*(4), 231–236.

Seed, H. B. (1973). Analysis of the slides in the San Fernando Dams during the earthquake of Feb. 9, 1971: Report to State of California Department of Water Resources, Los Angeles Department of Water and Power, National Science Foundation. College of Engineering, University of California.

Serff, N. (1976). Earthquake induced deformations of earth dams. College of Engineering, University of California.

Shi, G. (2002). Single and multiple block limit equilibrium of key block method and discontinuous deformation analysis. *Proceedings of the 5th international conference on analysis of discontinuous deformation* (pp. 3–43). AA Balkema: Rotterdam.

Shi, G.-H., & Goodman, R. E. (1985). Two dimensional discontinuous deformation analysis. *International Journal for Numerical and Analytical Methods in Geomechanics, 9,* 541–556.

Shi, G.-H., & Goodman, R. E. (1989). Generalization of two-dimensional discontinuous deformation analysis for forward modelling. *International Journal for Numerical and Analyrucal Methods in Geomechanics, 13,* 359–380.

Sosio, R., Crosta, G. B., Chen, J. H., & Hungr, O. (2012). Modelling rock avalanche propagation onto glaciers. *Quaternary Science Reviews, 47,* 23–40.

Taiebat, M., Kaynia, A. M., & Dafalias, Y. F. (2011). Application of an Anisotropic Constitutive Model for Structured Clay to Seismic Slope Stability. *Journal of Geotechnical and Geoenvironmental Engineering, 137*(5), 492.

Tang, C., Zhu, J., Qi, X., & Ding, J. (2011). Landslides induced by the Wenchuan earthquake and the subsequent strong rainfall event: A case study in the Beichuan area of China. *Engineering Geology, 122*(1–2), 22–33.

Tsesarsky, M., Hatzor, Y. H., & Sitar, N. (2005). Dynamic displacement of a block on an inclined plane: Analytical, experimental and DDA results. *Rock Mechanics and Rock Engineering, 38* (2), 153–167.

Wu, J.-H. (2003). *Numerical analysis of discontinuous rock masses using discontinuous deformation analysis*. Kyoto, Japan: Kyoto University.

Wu, J.-H. (2007). Applying discontinuous deformation analysis to assess the constrained area of the unstable Chiu-fen-erh-shan landslide slope. *International journal for numerical and analytical methods in geomechanics, 31*(5), 649–666.

Wu, J.-H. (2010). Seismic landslide simulations in discontinuous deformation analysis. *Computers and Geotechnics, 37*(5), 594–601.

Wu, J.-H., & Chen, C.-H. (2011). Application of DDA to simulate characteristics of the Tsaoling landslide. *Computers and Geotechnics, 38*(5), 741–750.

Wu, J.-H., & Tsai, P.-H. (2011). New dynamic procedure for back-calculating the shear strength parameters of large landslides. Engineering Geology.

Wu, J.-H., Ohnishi, Y., & Nishiyama, S. (2005). A development of the discontinuous deformation analysis for rock fall analysis. *International journal for numerical and analytical methods in geomechanics, 29*(10), 971–988.

Wu, J., Lin, J., & Chen, C. (2009). Dynamic discrete analysis of an earthquake-induced large-scale landslide. *International Journal of Rock Mechanics and Mining Sciences, 46*(2), 397–407.

Yagoda-Biran, G., & Hatzor, Y. H. (2010). Constraining paleo PGA values by numerical analysis of overturned columns. *Earthquake Engineering & Structural Dynamics, 39*(4), 463–472.

Yin, Y., Wang, F., & Sun, P. (2009). Landslide hazards triggered by the 2008 Wenchuan earthquake, Sichuan, China. *Landslides, 6*(2), 139–152.

Yin, Y., Yu, C., Wang, J., Wang, M., Liu, B., Song, Y., et al. (2011). Remote sensing research on Daguangbao gigantic rockslide triggered by Wenchuan earthquake. *Journal of Engineering Geology, 19*(5), 674–684. (In Chinese).

Zhang, Y., Chen, G., & Zheng, L. (2013a). Detailed investigation of near-fault earthquake loading induced displacement of sliding system by the discontinuous deformation analysis (DDA). In Proc. of the 11th International Conference of Analysis of Discontinuous Deformation. NRC, Fukuoka, Japan.

Zhang, Y., Chen, G., Zen, K., & Kasama, K. (2011). High-speed starting mechanism of rock avalanches induced by earthquake. *International conference on advances in Geotechnical Engineering* (pp. 803–809). Australia: Perth.

Zhang, Y., Chen, G., Zheng, L., Li, Y., & Zhuang, X. (2013b). Effects of geometries on three-dimensional slope stability. *Canadian Geotechnical Journal, 50*(3), 233–249.

Zhang, Y., Chen, G., Wang, B., & Li, L. (2013c). An analytical method to evaluate the effect of a turning corner on 3D slope stability. *Computers and Geotechnics, 53,* 40–45.

Zheng, L. (2010). *Development of new models for landslide simulation based on discontinuous deformation analysis*. Fukuoka, Japan: Kyushu University.

Zheng, Y. R., Tang, X., Zhao, S., Deng, C., & Lei, W. (2009). Strength reduction and step-loading finite element approaches in geotechnical engineering. *Journal of Rock Mechanics and Geotechnical Engineering, 1*(1), 21–30.

Zienkiewicz, O. C., Humpheson, C., & Lewis, R. W. (1975). Associated and non-associated visco-plasticity and plasticity in soil mechanics. *Géotechnique, 25*(4), 671–689.

Chapter 7
Conclusions and Future Studies

Abstract This chapter summarizes and concludes the results and achievements of the study. Problems are also highlighted for future studies.

Keywords Conclusions · Future studies

7.1 Conclusions

A large number of landslides can be induced by a strong earthquake, and they can cause very serious damage to both lives and properties. In order to mitigate the landslide disasters, it is important and necessary to analyze the stability of a slope and movement behaviors of a potential landslide under seismic loadings. Therefore, this study focuses on analysis of (1) slope stability and (2) landslide movement behaviors.

For the slope stability analysis, two major approaches are widely used. One is to calculate the pseudo-static safety factor of a slope. The other is to calculate earthquake-induced permanent displacement. In this monograph, a new analytical method and an extensional numerical simulation method were presented for slope stability analysis based on both shear and tensile failure modes under seismic loadings. And a new dynamic sliding block method for calculating the earthquake-induced permanent displacement of slope was presented.

In the study of landslide movement behaviors, long run-out, one of the major behaviors of earthquake-induced landslides, is discussed. Since the mechanism is still not clear, the MAM is extended based on the so-called trampoline effect. Also, a practical numerical simulation program is developed to clarify movement behaviors of earthquake-induced landslides by practical simulations.

The following major conclusions can be drawn:

1. A larger number of analysis examples show that the influence of tension failure on slope stability is significant on FOS. FOS from tension-shear failure mode is smaller than that from traditional single-shear failure mode. FOS could not be correct if tensile failure is ignored in the case of seismic loading.

Y. Zhang, *Earthquake-Induced Landslides*, Springer Natural Hazards,
DOI 10.1007/978-981-10-2935-6_7

2. Results also show that tension failure has a significant influence on the failure surface of a slope. The shape of failure surface of seismic slope under tension-shear failure mechanism is obviously different from that of the traditional shear-only failure mechanism. Failure surface of a seismic slope considering tension-shear failure mechanism contains two segments and is shallower than that just considering the shear failure while ignoring the influence of tension failure.
3. Compared to the widely acceptable influence of horizontal seismic loading on slope stability, the vertical seismic loading should be taken more attention, especially for some slopes under extreme seismic loadings.
4. The $FLAC^{3D}$ program can take the influence of tension failure into consideration, so as to make the analysis more reasonable for practical application. Results from a studied case show that $FLAC^{3D}$ gives almost the same FOS and failure surface compared to the limit analysis.
5. Both limit analysis and $FLAC^{3D}$ can be used to determine reasonable FOS which is smaller than that obtained from traditional single-shear failure mechanism. The widespread use of this mechanism should now be seriously considered to replace the traditional failure mechanism.
6. The MAM was extended for landslide run-out analysis. The effects of seismic loading on the landslide run-out can be taken account into this model. The MAM can be used to accurately estimate the long run-out of landslide induced by earthquake.
7. A practical numerical simulation program is developed based on the MAM incorporating with the extended DDA. Trampoline effects induced by extreme seismic loading can be successfully reproduced by the new seismic DDA program. Some case studies of large-scale landslides induced by the 2008 Wenchuan earthquake show that the proposed new long run-out model is reasonable and applicable and the movement behavior of earthquake-induced landslides can be analyzed by using the numerical simulation program.
8. The Daguangbao landslide, the largest landslide induced by the 2008 Wenchuan earthquake, was analyzed by numerical simulation methods for the states of initiation and run-out. The simulation results show that the vertical component of seismic loading may play an important role on both the initiation and run-out analysis since larger tension failure and trampoline effects may be induced by the vertical seismic force, which is generally ignored up to now.

7.2 Future Studies

1. More investigations and shaking table tests should be carried out to validate the effects of tension failure on the seismic slope stability. At present, most studies about the tension failure focus on the qualitative analysis, and the presented

method in this monograph carried out a preliminary quantitative analysis. Further quantitative analysis should be continued.

2. Both natural and constructed slopes exhibit a complex configuration and a 3D state. Therefore, 3D analysis method for seismic slope stability based on tension-shear failure mechanism should be presented. And other external forces (e.g., surcharge, pore water press, anchor force) should be considered.
3. The coseismic permanent displacement of slope should be further studied using a complex model rather than a simple single-sliding block. In addition, rigorous elastic-plastic model should be presented to take the more realistic feature of sliding mass rather than rigid-plastic model.
4. The presented long run-out of landslide should be validated by experiments and other numerical simulation methods. In order to take the 3D topography into consideration of landslide run-out, the developed practical numerical simulation program should be extended to 3D.

Appendix A

See Appendix Table A.1.

Y. Zhang, *Earthquake-Induced Landslides*, Springer Natural Hazards,
DOI 10.1007/978-981-10-2935-6

Table A.1 List of historical earthquakes causing substantial lanslideing

No.	Date D/M/Y	Earthquake	Country	Magnitude		Dep. (km)	Area of landslides (km^2)	Landslides		References
				Ms	Mw			Order	Number	
1	18/04/1906	San Francisco, California	USA	8.25–8.3	7.9	≤ 20	36,342	4–5		Keefer (1984)
2	31/03/1917	Daguan, Yunan	China	6.75					>40	Sun and Cai (1997), Qiao and Pu (1992)
3	16/12/1920	Haiyuan, Gansu	China	8.5	7.8	25	5600		657	Keefer (1984, 2000), Chen et al. (2005)
4	24/03/1923	Luhuo, Sichuan	China	7.25					120	Qiao and Pu (1992)
5	23/05/1927	Gulang, Gansu	China	8.0*			2100		61	Yamada (2013), Chen et al. (2005)
6	15/01/1934	Bihar	India-Nepal	8.3	8.1	15	128,453	4–5		Keefer (1984)
7	27/04/1936	Mabian, Sichuan	China	6.75					>30	Qiao and Pu (1992)
8	19/05/1940	Imperial Valley, California	USA	7.1		16	1494	3–4		Keefer (1984)
9	23/06/1946	Vancouver Island	Canada	7.2–7.3		30	22,893	3		Keefer (1984)
10	28/06/1948	Fukui	Japan	7.25–7.3*		14–33		3–4		Keefer (1984)
11	13/04/1949	Puget Sound, Washington	USA	7		70	12,538	2		Keefer (1984)
12	10/07/1949	Khait	Tajikistan	7.4*					A large number	Evans et al. (2009)
13	25/05/1948	Litang, Sichuan	China	7.3					>70	http://baike.baidu.com/view/6162157.htm
14	15/08/1950	Assam	India	8.6–8.7	8.6	14	50,970	5–>5		Keefer (1984)
15	14/04/1955	Kangding, Sichuan	China	7.5					>30	Tang et al. (2001)

(continued)

Table A.1 (continued)

No.	Date D/M/Y	Earthquake	Country	Magnitude		Dep. (km)	Area of landslides (km^2)	Landslides		References
				Ms	Mw			Order	Number	
16	22/03/1957	Daly City, California	USA	5.3		7–10	9	2	23	Keefer (1984, 2002), Bonilla (1960)
17	10/07/1958	Southest Alaska	USA	7.9	7.7	15	66,019	3–4		Keefer (1984)
18	18/08/1959	Hebgen Lake, Montana	USA	7.1*		10–12	4560	3–4		Keefer (1984)
19	28/03/1964	Alaska	USA	8.3–8.4	9.2	20–50	273,684	4–5		Keefer (1984, 2000)
20	16/06/1964	Niigata	Japan	7.3		40		3–4		Keefer (1984)
21	29/04/1965	Puget Sound, Washington	USA	6.5		58–63	4624	2		Keefer (1984)
22	28/06/1966	Parkfield-Cholame, California	USA	6.2		4–10	341	2–3		Keefer (1984)
23	23/05/1968	Inangahua	New Zealand	7.1		10		2–3		Keefer (1984), Hancox et al. (2001)
24	05/01/1970	Tonghai, Yunan	China	7			85		>40	Sun and Cai (1997), Tang et al. (2001)
25	31/05/1970	Peru	Peru	7.8	7.9	35–43	30,174	4–5		Keefer (1984), Harp et al. (2011), Plafker et al. (1971), Tang et al. (1997)
26	31/10/1970	Madang	Papua New Guines	7.1		41	10,211	4		Keefer (1984)
27	09/02/1971	San Francisco, California	USA	6.5		8–13	3392	3–4		Keefer (1984)
28	26/02/1973	Luhuo, Sichuan	China	7.9			90		139	Yamada (2013)
29	26/04/1973	Honomu, Hawaii	USA	6.1		41–50	881	2–3		Keefer (1984)

(continued)

Table A.1 (continued)

No.	Date D/M/Y	Earthquake	Country	Magnitude		Dep. (km)	Area of landslides (km^2)	Landslides		References
				Ms	Mw			Order	Number	
30	11/05/1974	Zhaotong, Yunnan	China	7.1*					79	Yamada (2013), Sun and Cai (1997), Tang et al. (2001), Qiao and Pu (1992)
31	28/10/1974	Indus Kohistan	Pakistan	6.2		12	779	4	4	Keefer (1984)
32	29/11/1975	Kilauea, Hawaii	USA	7.1		5	6615	3	3	Keefer (1984)
33	04/02/1976	Guatemala	Guatemala	7.5	7.5	5	16,840	5	~50,000	Keefer (1984, 2002), Harp et al. (1981, 2011)
34	29/05/1976	Longling, Yunnan	China	7.3–7.4					151	Yamada (2013), Wang and Nie (2009), Chen and Zhao (1979)
35	06/05/1976	Friuli	Italy	6.3–6.5		8–26	2106	4–5		Keefer (1984)
36	29/05/1976	Longling, Yunnan	China	7.4					141	Tang et al. (2001)
37	11/07/1976	Darien	Panama	7		3		3–4		Keefer (1984)
38	27/07/1976	Tangshan	China	7.7–8.0		12–16	28,229	3–4		Keefer (1984)
39	16, 23/08/1976	Songpan, Pingwu, Nanping and Wenxian, Sichuan	China	7.2					>130	Yamada (2013), Wang and Nie (2009)
40	21/03/1977	Khurgu	Iran	6.9		29		3–4		Keefer (1984)
41	23/11/1977	San Juan Province	Argentina	7.4		17	9829	4–5		Keefer (1984)
42	14/01/1978	Izu-Oshima Kinkai	Japan	6.8	7	4	291	3–4		Keefer (1984), Harp et al. (2011)
43	12/06/1978	Miyagi-ken-oki	Japan	7.4		30	4300	3–4		Keefer (1984)
44	13/08/1978	Sants Barbara, California	USA	5.6		13	312	2–3		Keefer (1984)

(continued)

Table A.1 (continued)

No.	Date D/M/Y	Earthquake	Country	Magnitude		Dep. (km)	Area of landslides (km^2)	Landslides		References
				Ms	Mw			Order	Number	
45	15/03/1979	Homestead Valley, California	USA	5.2*		≤ 4		1–2		Keefer (1984)
46	06/08/1979	Coyote Lake, California	USA	5.4		10	253	2		Keefer (1984)
47	24/01/1980	Mount Diablo, California	USA	5.8		8	492	2	103	Keefer (1984, 2002), Wilson and Keefer (1983, 1989)
48	24/05/1980	Mammoth Lakes, California	USA	6.1	6.2	8	1201	4	5253	Keefer (1984, 2002), Harp et al. (1984, 2011), Keefer and Wilson (1989)
49	23/11/1980	Irpinia	Italy	6.9	6.9	0	13,000	2		Rodriguez et al. (1999), Wasowski et al. (2002)
50	02, 09/05/1983	Coalinga	USA	6.7	6.5	7	650	4	9389	Rodriguez et al. (1999), Harp et al. (2011), Harp and Keefer (1990, 2002), Keefer and Wilson (1989)
51	28/10/1983	Borah Peak	USA	7.3	6.9	10	4200	2		Rodriguez et al. (1999)
52	14/09/1984	Nagoken-Seibu	Japan	6.4	6.2	18	3500	2		Rodriguez et al. (1999)
53	03/03/1985	Valparaiso	Chile	7.8	8.1	43	17,000			Rodriguez et al. (1999)
54	13/09/1986	Kalamato	Greece	5.8	6	28	70	2		Rodriguez et al. (1999)
55	10/10/1986	San Salvador	El Salvador	5.4	5.7	12	380	4	>216	Rodriguez et al. (1999), Keefer (2002), Rymer (1987), Rymer and White (1989)
56	07/01/1987	Diebu	China	5.5	5.3	15	280			Rodriguez et al. (1999)

(continued)

Table A.1 (continued)

No.	Date D/M/Y	Earthquake	Country	Magnitude		Dep. (km)	Area of landslides (km^2)	Landslides		References
				Ms	Mw			Order	Number	
57	02/03/1987	Edgecumbe	New Zealand	6.6	6.5	6	380	2		Rodriguez et al. (1999), Hancox et al. (2001)
58	06/03/1987	El Napo	Ecuador	6.9	7.1	5–15	2500			Rodriguez et al. (1999)
59	01, 04/10/1987	Whittier Narrows	USA	5.8	6.1	17	4200			Rodriguez et al. (1999)
60	24/11/1987	Superstition Hills	USA	6.6	6.6	1.9	3300			Rodriguez et al. (1999)
61	21/08/1988	Nepal	Nepal	6.6	6.8	65	90			Rodriguez et al. (1999)
62	16/10/1988	Killini	Greece	5.6	5.9	25	40	2		Rodriguez et al. (1999), Papadopoulos and Plessa (2000), Papadopoulos and Profis (1990)
63	20/10/1988	Lancang, Gengma, Yunan	China	7.6/7.2					160	http://www.csi.ac.cn/
64	06/11/1988	Lancang, Yunnan	China	7.6					160	Yamada (2013), Wang and Nie (2009)
65	25/11/1988	Seguenay	Canada	5.8	5.8	28	45,000	2		Rodriguez et al. (1999)
66	07/12/1988	Spitak	Armenia	6.8	6.7	5	2200	4		Rodriguez et al. (1999)
67	23/01/1989	Soviet Tajik	Tajikistan	5.5		10	12	2		Rodriguez et al. (1999)
68	17/10/1989	Loma Prieta, California	USA	7.1	6.9	8	15,000	4	~1500	Rodriguez et al. (1999), Harp et al. (2011), Keefer (2000, 2002), Keefer and Manson (1998)
69	20/06/1990	Manjil	Iran	7.3	7.4	19	1000	3		Rodriguez et al. (1999)
70	16/07/1990	Luzon	Philippines	7.8	7.7	25	3000	3		Rodriguez et al. (1999)

(continued)

Table A.1 (continued)

No.	Date D/M/Y	Earthquake	Country	Magnitude		Dep. (km)	Area of landslides (km^2)	Landslides		References
				Ms	Mw			Order	Number	
71	22/04/1991	Valle de la Estrella	Costa Rica	7.6	7.5	21.5	2000	4		Rodriguez et al. (1999)
72	13,15/03/1992	Erzincan	Turkey	6.8	6.7	27	150	2		Rodriguez et al. (1999)
73	25, 26/04/1992	Cape Mendocino	USA	7.1	7	15	625			Rodriguez et al. (1999)
74	19/08/1992	Suusamyr	Kyrgytzstan	7.3	7	16.5	2500			Rodriguez et al. (1999)
75	17, 18/10/1992	Murindo	Colombia	6.7–7.3	6.7–7.4	10–14	9700			Rodriguez et al. (1999)
76	12/07/1993	Hokkaido-Nansei	Japan	7.6	7.8	17	100	2		Rodriguez et al. (1999), Harp et al. (2011)
77	10/08/1993	Ormond	New Zealand	6.2	6.2	39	5	2		Rodriguez et al. (1999)
78	10/08/1993	Fiorland	New Zealand	7	7	20	500	2		Rodriguez et al. (1999)
79	21/09/1993	Klamath Falls	USA	5.8	5.9	5	420			Rodriguez et al. (1999)
80	17/01/1994	Northridge	USA	6.8	6.7	18	10,000	5	>11,000	Harp and Jibson (1995), Rodriguez et al. (1999), Harp et al. (2011), Jibson et al. (2000), Pradel et al. (2005)
81	06/06/1994	Paez	Colombia	6.6	6.8	12	250	4		Rodriguez et al. (1999)
82	18/06/1994	Arthur's Pass	New Zealand	7.1	6.8	4	85			Rodriguez et al. (1999)
83	17/01/1995	Hyogu-Ken Nanbu (Kobe)	Japan	6.8	6.9	22	910	3	674–747	Rodriguez et al. (1999), Harp et al. (2011), Keefer (2002)

(continued)

Table A.1 (continued)

No.	Date D/M/Y	Earthquake	Country	Magnitude		Dep. (km)	Area of landslides (km^2)	Landslides		References
				Ms	Mw			Order	Number	
84	19/01/1995	Tauramena	Colombia	6.6	6.5	20	4550			Rodriguez et al. (1999)
85	29/05/1995	Arthur's Pass	New Zealand	6.5		4	85	2		Rodriguez et al. (1999)
86	03/02/1996	Lijiang, Yunan	China	7		10			>450	Tang et al. (1997, 2001)
87	26/09/1997	Umbria-Marche	Italy	5.5–5.9	6	3–7			100–124	Rodriguez et al. (1999), Keefer (2002), Bozzano (1998), Esposito et al. (2000)
88	12/04/1998	Upper Soča River basin	Slovenia	ML=5.6					>100	Mikoš et al. (2006)
89	21/09/1999	Chi-Chi, Nantou County	Taiwan, China	7.3	7.7	7	10,000		22,000	Yamada (2013), Harp et al. (2011), Keefer (2002), Hung (2000), Luzi et al. (2000), Shou et al. (2011)
90	22/06/2002	Avaj	Iran		6.5	5–10	>3600		59	Mahdavifar et al. (2006)
91	21/01/2003	Tecoman (Colima)	Mexico		7.6	10–24			Several hundred	Keefer et al. (2006)
92	12/07/2004	Upper Soča River basin	Slovenia	ML=4.9					50	Mikoš et al. (2006)
93	23/10/2004	Niigata, Cyuetsu	Japan	Mj=6.8			1316		3791	Yamada (2013), Harp et al. (2011), Deng et al. (2010, 2011)

(continued)

Table A.1 (continued)

No.	Date D/M/Y	Earthquake	Country	Magnitude		Dep. (km)	Area of landslides (km^2)	Landslides		References
				Ms	Mw			Order	Number	
94	12/05/2008	Wenchuan	China	8	7.9	19	>100,000		>60,000	Yamada (2013), Qiao et al. (2009), Ouimet (2010)
95	14/06/2008	Iwate and Miyagi	Japan	7.2			400		3000	Yamada (2013), Harp et al. (2011), Yagi et al. (2009), Miyagi et al. (2011)
96	10/01/2010	Humboldt County, California	USA		6.5				11	Storesund et al. (2010)
97	12/01/2010	Haiti	Haiti	7*					4000–5000	Jibson and Harp (2011) (USGS)
98	22/2/2011	Christchurch	New Zealand		6.3	Shallow	150		140	Wilkinson et al. (2013), Hancox and Perrin (2011), Carydis (2012)
99	11/03/2011	Tohoku	Japan		9				2000	Yamada (2013), Miyagi et al. (2011)

Note Number with the mark *means magnitude of the earthquake is M, not special for Ms or Mw

Appendix B

See Appendix Table B.1.

Table B.1 List of near-fault earthquake records utilized as excitations in this study

	Data (year/month/day) Earthquake, Country Magnitude	No. of STA	DF (km)	Station/Record name	PGA (cm/s^2)	PGA_v/ PGA_h
1	1971/2/9 San Fernando, California, USA Ms=6.5, Mw=6.6	1	3.5	Pacoima Dam-164	1148.06	0.61
				Pacoima Dam-254	1054.95	0.66
				Pacoima Dam-down	695.97	1.00
2	1976/5/17 Gazli, USSR Ms=7.0	2	5.5	Karakyr-0°	596.70	2.08
				Karakyr-90°	703.92	1.76
				Karakyr-up	1240.00	1.00
3	1978/9/16 Tabas, Iran Ms=7.4	3	2.1	Tabas-LN	819.93	0.82
				Tabas-TR	835.58	0.81
				Tabas-UP	675.42	1.00
4	1979/10/15 Imperial Valley, California, USA Ms=6.9, Mw=6.5	4	3.1	5028 El Centro Array #7-140°	331.28	1.40
				5028 El Centro Array #7-230°	450.32	1.03
				5028 El Centro Array #7-up	462.40	1.00
		4*	0.07	5155 EC Meloland Overpass FF-0°	0.314 g	
				5155 EC Meloland Overpass FF-270°	0.296 g	
				5155 EC Meloland Overpass FF-up	0.248 g	
		5	3.3	6618 Agrarias-3	351.40	2.53
				6618 Agrarias-273	229.90	3.87
				6618 Agrarias-down	889.40	1.00

(continued)

Y. Zhang, *Earthquake-Induced Landslides*, Springer Natural Hazards,
DOI 10.1007/978-981-10-2935-6

Table B.1 (continued)

	Data (year/month/day) Earthquake, Country Magnitude	No. of STA	DF (km)	Station/Record name	PGA (cm/s²)	PGAᵥ/ PGAₕ
		6	3.5	5158 El Centro Array #6-140	332.44	4.84
				5158 El Centro Array #6-230	429.18	3.75
				5158 El Centro Array #6-up	1610.63	1.00
		7	4.4	5054 Bonds Corner-140	575.73	0.60
				5054 Bonds Corner-230	770.42	0.45
				5054 Bonds Corner-up	347.68	1.00
		8	5.2	0952 El Centro Array #5-140	539.78	0.87
				0952 El Centro Array #5-230	360.37	1.30
				0952 El Centro Array #5-up	469.30	1.00
		9	8.3	0955 El Centro Array #4-140	480.76	0.43
				0955 El Centro Array #4-230	349.18	0.59
				0955 El Centro Array -up	204.42	1.00
5	1989/10/17 Loma Prieta, California, USA Ms=7.1, Mw=6.8	10	3.9	57007 Corralitos-0°	631.76	0.71
				57007 Corralitos-90°	469.90	0.95
				57007 Corralitos-up	446.36	1.00
		11	8.1	Saratoga Aloha Avenue-0°	502.27	0.76
				Saratoga Aloha Avenue-90°	317.84	1.20
				Saratoga Aloha Avenue-up	381.61	1.00
		12	9.6	47379 Gilroy Array #1-0°	403.19	0.51
				47379 Gilroy Array #1-90°	464.01	0.44
				47379 Gilroy Array #1-up	205.03	1.00

(continued)

Table B.1 (continued)

	Data (year/month/day) Earthquake, Country Magnitude	No. of STA	DF (km)	Station/Record name	PGA (cm/s^2)	PGA_v/ PGA_h
		13	10.0	47006 Gilroy-Gavilan Coll-67°	350.22	0.54
				47006 Gilroy-Gavilan Coll-337°	318.83	0.59
				47006 Gilroy-Gavilan Coll-up-up	187.37	1.00
6	1992/3/13 Erzincan, Turkey Ms=6.9	14	2.0	ERZINCAN (No.95)-EW	486.58	0.50
				ERZINCAN (No.95)-NS	505.22	0.48
				ERZINCAN (No.95)-UD	243.29	1.00
7	1992/6/28 Landers, California, USA Ms=7.8, Mw=7.3	15	2.0	Lucerne Valley-0	798.00	
				Lucerne Valley-90	716.80	
8	1994/1/17 Northridge, California, USA Ms=6.8, Mw=6.7	16	5.0*	24436 Tarzana, Cedar Hill-90	1745.20	0.59
				24436 Tarzana, Cedar Hill-360	971.19	1.06
				24436 Tarzana, Cedar Hill-up	1028.09	1.00
9	1994/1/17 Northridge, California, USA Ms=6.8, Mw=6.7	17	8.6	Los Angeles Dam-64	317.60	1.00
				Los Angeles Dam-334	419.10	0.76
				Los Angeles Dam-up	316.50	1.00
		18	8.6	77 Rinaldi Receiving Sta-228	825.50	1.01
				77 Rinaldi Receiving Sta-318	471.00	1.76
				77 Rinaldi Receiving Sta-UP	830.00	1.00
		19	8.6	0655 Jensen Filter Plant-22	637.90	1.01
				0655 Jensen Filter Plant-292	825.70	0.78
				0655 Jensen Filter Plant-up	644.50	1.00

(continued)

Table B.1 (continued)

	Data (year/month/day) Earthquake, Country Magnitude	No. of STA	DF (km)	Station/Record name	PGA (cm/s²)	PGA_v/ PGA_h
		20	8.7	74 Sylmar—Converter Sta-52	592.60	1.30
				74 Sylmar—Converter Sta-142	739.50	1.04
				74 Sylmar—Converter Sta-up	770.80	1.00
		21	9.5	0637 Sepulveda VA-270	738.20	0.63
				0637 Sepulveda VA-360	922.70	0.51
				0637 Sepulveda VA-up	466.50	1.00
		22	9.5	Arleta—Nordhoff Fire Sta-=90	337.30	1.61
				Arleta—Nordhoff Fire Sta-360	302.00	1.79
				Arleta—Nordhoff Fire Sta-up	541.40	1.00
10	1995/1/16 Kobe, Japan Mw=6.9	23	0.3	TAKARAZU-0	680.03	0.62
				TAKARAZU-90	680.18	0.62
				TAKARAZU-up	424.97	1.00
		24	0.9	KOBE UNIVERSITY-0	284.47	1.31
				KOBE UNIVERSITY-90	304.46	1.22
				KOBE UNIVERSITY-up	372.90	1.00
		25	1.0	KJMA-0	805.45	0.42
				KJMA-90	586.94	0.57
				KJMA-up	336.13	1.00
		26	1.5	TAKATORI-0	599.59	0.44
				TAKATORI-90	603.61	0.44
				TAKATORI-up	266.37	1.00
		27	3.3	KOBE PORT ISLAND-0	308.73	1.79
				KOBE PORT ISLAND-90	272.35	2.02
				KOBE PORT ISLAND-up	551.39	1.00
		28	7.1	NISHI-AKASHI-0	499.49	0.73
				NISHI-AKASHI-90	493.03	0.74
				NISHI-AKASHI-up	363.56	1.00

(continued)

Table B.1 (continued)

	Data (year/month/day) Earthquake, Country Magnitude	No. of STA	DF (km)	Station/Record name	PGA (cm/s^2)	PGA_v/ PGA_h
11	1999/8/17 Kocaeli, Turkey Ms=7.4, Mw=7.8	29	3.1	Sakarya-90°	368.86	0.69
				Sakarya-up	254.08	1.00
		30	4.8	Yarimca-60°	262.91	0.90
				Yarimca-330°	342.37	0.69
				Yarimca-up	237.40	1.00
12	1999/9/21 Chi-Chi, Taiwan-China Ms=7.6, Mw=7.6	31	1.2	TCU102-0°	167.50	1.03
				TCU10 2-90°	297.10	0.58
				TCU102-UP	172.40	1.00
		32	1.8	TCU052-0°	437.20	0.45
				TCU052-90°	352.20	0.55
				TCU052-UP	194.60	1.00
		33	1.9	TCU101-0°	251.20	0.63
				TCU101-90°	207.30	0.77
				TCU101-UP	158.80	1.00
		34	2.2	TCU129-0°	611.30	0.55
				TCU129-90°	983.90	0.34
				TCU129-UP	335.30	1.00
13	1999/9/21 Chi-Chi, Taiwan-China Ms=7.6, Mw=7.6	35	2.2	WNT-0°	606.60	0.52
				WNT-90°	937.30	0.33
				WNT-UP	312.40	1.00
		36	2.4	TCU103-0°	149.00	0.96
				TCU103-90°	126.50	1.14
				TCU103-UP	143.70	1.00
		37	2.5	TCU065-0°	563.70	0.46
				TCU065-90°	772.70	0.33
				TCU065-UP	257.60	1.00
		38	3.0	TCU068-0°	353.10	1.47
				TCU068-90°	494.40	1.05
				TCU068-UP	520.20	1.00
		39	3.1	CHY080-0°	836.80	0.85
				CHY080-90°	792.80	0.90
				CHY080-UP	713.30	1.00
		40	3.2	TCU076-0°	421.20	0.65
				TCU076-90°	336.10	0.82
				TCU076-UP	274.90	1.00

(continued)

Table B.1 (continued)

Data (year/month/day) Earthquake, Country Magnitude	No. of STA	DF (km)	Station/Record name	PGA (cm/s^2)	PGA_v/ PGA_h
	41	3.3	TCU049-0°	238.40	0.75
			TCU049-90°	270.50	0.66
			TCU049-UP	179.30	1.00
	42	3.4	TCU075-0°	256.60	0.87
			TCU075-90°	324.80	0.69
			TCU075-UP	223.80	1.00
	43	3.4	TCU087-0°	111.30	0.82
			TCU087-90°	119.00	0.77
			TCU087-UP	91.30	1.00
	44	4.5	TCU-0°	187.20	0.63
			TCU-90°	200.50	0.59
			TCU-UP	117.90	1.00
	45	4.5	TCU082-0°	182.10	0.71
			TCU082-90°	219.50	0.59
			TCU082-UP	129.10	1.00
	46	4.6	TCU054-0°	189.20	0.71
			TCU054-90°	142.60	0.94
			TCU054-UP	133.40	1.00
	47	4.9	TCU071-0°	639.60	0.65
			TCU071-90°	518.40	0.80
			TCU071-UP	416.00	1.00
	48	5.5	TCU053-0°	131.50	0.92
			TCU053-90°	224.30	0.54
			TCU053-UP	120.80	1.00
	49	5.6	TCU055-0°	208.10	0.73
			TCU055-90°	252.40	0.60
			TCU055-UP	152.30	1.00
	50	7.0	TCU051-0°	230.00	0.47
			TCU051-90°	157.00	0.69
			TCU051-UP	109.00	1.00
	51	7.5	TCU136-0°	170.50	0.65
			TCU136-90°	166.80	0.67
			TCU136-UP	111.20	1.00
	52	7.9	TCU072-0°	360.10	0.77
			TCU072-90°	466.90	0.59
			TCU072-UP	276.00	1.00

(continued)

Table B.1 (continued)

	Data (year/month/day) Earthquake, Country Magnitude	No. of STA	DF (km)	Station/Record name	PGA (cm/s^2)	PGA_v/ PGA_h
14	1999/9/21 Chi-Chi, Taiwan-China Ms=7.6, Mw=7.6	53	8.1	TCU060-0°	99.50	0.87
				TCU060-90°	197.00	0.44
				TCU060-UP	86.70	1.00
		54	8.3	TCU078-0°	301.80	0.56
				TCU078-90°	433.60	0.39
				TCU078-UP	170.40	1.00
		55	8.7	CHY028-0°	693.20	0.50
				CHY028-90°	582.60	0.59
				CHY028-UP	343.30	1.00
		56	8.9	TCU050-0°	127.50	0.67
				TCU050-90°	142.70	0.60
				TCU050-UP	85.90	1.00
		57	9.1	NSY-0°	114.10	0.73
				NSY-90°	117.90	0.71
				NSY-UP	83.60	1.00
		58	9.1	TCU128-0°	163.50	0.55
				TCU128-90°	140.40	0.64
				TCU128-UP	90.50	1.00
		59	9.2	TCU122-0°	255.70	0.91
				TCU122-90°	207.70	1.12
				TCU122-UP	232.50	1.00
		60	9.3	CHY024-0°	160.40	0.88
				CHY024-90°	276.20	0.51
				CHY024-UP	141.50	1.00
		61	9.8	TCU056-0°	139.40	0.83
				TCU056-90°	152.90	0.76
				TCU056-UP	116.30	1.00
		62	9.9	TCU120-0°	193.00	0.86
				TCU120-90°	222.50	0.75
				TCU120-UP	165.90	1.00
15	1999/11/12 Duzce, Turkey Ms=7.5, Mw=7.2	63	6.6	Duzce-180°	341.39	1.03
				Duzce-270°	524.835	0.67
				Duzce-up	350.22	1.00
		64	12.0	Bolu-0°	714.17	0.28
				Bolu-90°	806.382	0.25
				Bolu-up	199.14	1.00

(continued)

Table B.1 (continued)

	Data (year/month/day) Earthquake, Country Magnitude	No. of STA	DF (km)	Station/Record name	PGA (cm/s^2)	PGA$_v$/ PGA$_h$
16	2008/5/12 Wenchuan, Sichuan, China Ms=8.0, Mw=7.9	65	4.0	51MZQ-EW	824.12	0.76
				51MZQ-NS	802.71	0.78
				51MZQ-UD	622.91	1.00
17	2008/6/14 Iwate-Miyagi, Japan M=7.2	66	3.0	IWTH25-EW(S)	1432.59	2.70
				IWTH25-NS(S)	1143.23	3.38
				IWTH25-UD(S)	3866.01	1.00
		67		IWTH25-EW(D)	747.92	0.91
				IWTH25-NS(D)	1036.19	0.66
				IWTH25-UD(D)	680.76	1.00
18	2008/7/24 Iwate-Miyagi, Japan M=6.8	68	8.0	IWTH03-EW(S)	548.52	1.05
				IWTH03-NS(S)	475.03	0.85
				IWTH03-UD(S)	578.31	1.00
		69		IWTH03-EW(D)	93.09	0.53
				IWTH03-NS(D)	80.47	0.61
				IWTH03-UD(D)	49.30	1.00

EQ Earthquake, *STA* Station, *DF* Closest distance to fault, *SSR* Union of Soviet Socialist Republic. Data from Pacific earthquake engineering research center: NGA Database, COSMOS (Consortium of Organizations for Strong-Motion Observation Systems), and Kyoshin net (K-NET)

Bibliography

Bonilla, M. G. (1960). *Landslides in the San Francisco South quadrangle*. California: US Department of the Interior, Geological Survey.

Bozzano, E (1998). La crisi della morte. Armenia.

Carydis, P., Lekkas, E., & Kritikos, T. (2012). The Mw 7.1 September 4 2010 and Mw 6.3 February 22 2011, New Zealand Eqs. Comparison of EMS1998 and ESI2007 Data. In 15 wcee. Lisboa.

Chen, L., & Zhao, W. (1979). *Longling earthquake, 1976*. Beijing: Earthquake press.

Chen, Y., Shi, Y., Liu, H., & Lu, Y. (2005). Distribution characteristics and influencing factors analysis of seismic loess landslides. *Earthquake Research in China, 21*(2), 235–243.

Deng, J., Tsutsumi, Y., Kameya, H., & Koseki, J. (2010). A modified procedure to evaluate earthquake-induced displacement of slopes containing a weak layer. *Soils and Foundations, 50*(3), 413–420.

Deng, J., Kameya, H., Miyashita, Y., Kuwano, J., Kuwano, R., & Koseki, J. (2011). Study on a failed dip slope with a thin sandy layer in 2004 Niigata-ken Chuetsu earthquake. *Engineering Geology, 123*(4), 302–314.

Esposito, E., Porfido, S., Simonelli, A. L., Mastrolorenzo, G., & Iaccarino, G. (2000). Landslides and other surface effects induced by the 1997 Umbria–Marche seismic sequence. *Engineering Geology, 58*(3), 353–376.

Evans, S. G., Roberts, N. J., Ischuk, A., Delaney, K. B., Morozova, G. S., & Tutubalina, O. (2009). Landslides triggered by the 1949 Khait earthquake, Tajikistan, and associated loss of life. *Engineering Geology, 109*(3), 195–212.

Hancox, G. T., Perrin, N. D., & Dellow, G. D. (2002). Recent studies of historical earthquake-induced landsliding, ground damage, and MM intensity in New Zealand. *Bulletin of The New Zealand Society for Earthquake Engineering, 35*(2), 59–95.

Hancox, G., Massey, C., & Perrin, N. (2011). Landslides and related ground damage caused by the Mw 6.3 Christchurch earthquake of 22 February 2011. *New Zealand Geomechanics News, 81*, 53–67.

Harp, E., and Keefer, D. (1990). Landslides triggered by the earthquake. United States Geological Survey, Professional Paper;(USA), 1487.

Harp, E. L., & Jibson, R. W. (1995). *Inventory of landslides triggered by the 1994 Northridge*. US Geological Survey: California earthquake.

Harp, E. L., Wilson, R. C., & Wieczorek, G. F. (1981). Landslides from the February 4, 1976, Guatemala earthquake.

Harp, E. L., Tanaka, K., Sarmiento, J., & Keefer, D. (1984). Landslides from the May 25–27, 1980, Mammoth Lakes, California, earthquake sequence. US Geological Survey.

Harp, E. L., Keefer, D. K., Sato, H. P., & Yagi, H. (2011). Landslide inventories: The essential part of seismic landslide hazard analyses. *Engineering Geology, 122*(1–2), 9–21.

http://baike.baidu.com/view/6162157.htm 2013. Sichuan Litang earthquake.

Y. Zhang, *Earthquake-Induced Landslides*, Springer Natural Hazards,
DOI 10.1007/978-981-10-2935-6

http://www.csi.ac.cn/manage/html/4028861611c5c2ba0111c5c558b00001/_history/ymd/ymdnews/lc88110601.htm 2013. Lancang-gengma earthquake.

Hung, J.-J. (2000). Chi-Chi earthquake induced landslides in Taiwan. *Earthquake Engineering and Engineering Seismology, 2*(2), 25–33.

Jibson, R., & Harp, E. (2011). Field reconnaissance report of landslides triggered by the January 12, 2010, Haiti earthquake. *US Geological Survey Open-File Report, 1023,* 19.

Jibson, R., Harp, E. L., & Michael, A. J. (2000). A method for producing digital probabilistic seismic. *Engineering Geology, 58,* 271–289.

Keefer, D. K. (1984). Landslides caused by earthquakes. *Geological Society of America Bulletin, 95*(4), 406.

Keefer, D. K. (2000). Statistical analysis of an earthquake-induced landslide distribution—the 1989 Loma Prieta, California event. *Engineering Geology, 58*(3), 231–249.

Keefer, D. K. (2002). Investigating landslides caused by earthquakes—a historical review. *Surveys in Geophysics, 23,* 473–510.

Keefer, D., & Wilson, R. (1989). Predicting earthquake-induced landslides, with emphasis on arid and semi-arid environments. *Landslides in a semi-arid environment, 2,* 118–149.

Keefer, D.K., & Manson, M.W. (1998). Regional distribution and characteristics of landslides generated by the earthquake. US Geological Survey professional paper (1551C).

Keefer, D., Wartman, J., Navarroochoa, C., Rodriguezmarek, A., & Wieczorek, G. (2006). Landslides caused by the M 7.6 Tecomán, Mexico earthquake of January 21, 2003. *Engineering Geology, 86*(2–3), 183–197.

Luzi, L., Pergalani, F., & Terlien, M. (2000). Slope vulnerability to earthquakes at subregional scale, using probabilistic techniques and geographic information systems. *Engineering Geology, 58*(3), 313–336.

Mahdavifar, M. R., Solaymani, S., & Jafari, M. K. (2006). Landslides triggered by the Avaj, Iran earthquake of June 22, 2002. *Engineering Geology, 86*(2), 166–182.

Mikos, M., Fazarinc, R., & Ribicic, M. (2006). Sediment production and delivery from recent large landslides and earthquake-induced rock falls in the upper Soca river valley, Slovenia. *Engineering Geology, 86*(2), 198–210.

Miyagi, T., Higaki, D., Yagi, H., Doshida, S., Chiba, N., Umemura, J., et al. (2011). Reconnaissance report on landslide disasters in northeast Japan following the M 9 Tōhoku earthquake. *Landslides, 8*(3), 339–342.

Miyagi, T., Higaki, D., Yagi, H., Doshida, S., Chiba, N., Umemura, J., et al. (2011). Reconnaissance report on landslide disasters in northeast Japan following the M 9 Tōhoku earthquake. *Landslides, 8*(3), 339–342.

Ouimet, W. B. (2010). Landslides associated with the May 12, 2008 Wenchuan earthquake: Implications for the erosion and tectonic evolution of the Longmen Shan. *Tectonophysics, 491* (1), 244–252.

Papadopoulos, G. A., & Profis, T. (1990). Macroseismic observations related with the strong shock of October 16th, 1988, in NW Peloponnesus, Greece: The important role of the microzonation conditions. *Journal of Geodynamics, 12*(2), 217–231.

Papadopoulos, G. A., & Plessa, A. (2000). Magnitude–distance relations for earthquake-induced landslides in Greece. *Engineering Geology, 58*(3), 377–386.

Plafker, G., Ericksen, G. E., & Concha, J. F. (1971). Geological aspects of the May 31, 1970, Peru earthquake. *Bulletin of the Seismological Society of America, 61*(3), 543–578.

Pradel, D., Asce, M., Smith, P. M., Stewart, J. P., & Raad, G. (2005). Case history of landslide movement during the Northridge earthquake. *Journal of Geotechnical and Geoenvironmental Engineering* (November), 1360–1369.

Qiao, J., & Pu, X. (1992). A preliminary study on the distributive regulation of seismic landslide in Sichuan and Yuannan. *Journal of Seismological Research, 15*(4), 411–417.

Qiao, J., Pu, X., Wang, M., Yang, Z., Shi, L., & Tian, H. (2009). Ana lysis of d istr ibution character istics of landslide induced in Wenchuan earthquake and relevantmax imum epicenter distance. *Journal of Natural Disasters, 18*(5), 10–15.

Rodriguez, C. E., Bommer, J., & Chandler, R. J. (1999). Earthquake-induced landslides 1980-1997. *Soil Dynamics and Earthquake Engineering, 18,* 325–346.
Rymer, M. J. (1987). The San Salvador earthquake of October 10, 1986-geologic aspects. *Earthquake Spectra, 3*(3), 435–463.
Rymer, M., & White, R. (1989). Hazards in El Salvador from earthquake-induced landslides. In *Landslides: Extent and economic significance* (pp. 105–109). Balkema, Rotterdam.
Shou, K. J., Hong, C. Y., Wu, C. C., Hsu, H. Y., Fei, L. Y., Lee, J. F., & Wei, C. Y. (2011). Spatial and temporal analysis of landslides in central Taiwan after 1999 Chi-Chi earthquake. *Engineering Geology.*
Storesund, R., Dengler, L., Mahin, S., Collins, B. D., Hanshaw, M., Turner, F., & Welsh, K. (2010). M 6.5 earthquake offshore Northern California January 9, 2010.
Sun, C., & Cai, H. (1997). Developing and distributing characteristics of collapses and landslides during strong historic earthquake in China. *Journal of Natural Disasters, 6*(1), 25–30.
Tang, C., Huang, C., & Wang, Y. (1997). Lijiang earthquake and the induced rockfalls and slumps in Yunnan. *Journal of Natural Disasters, 6*(3), 76–84.
Tang, C., Zhu, J., & Zhang, X. (2001). GIS based earthquake triggered landslide hazard prediction. *Journal of Seismological Research, 24*(1), 73–81.
Wang, X., & Nie, G. (2009). Quick evaluation of distance harmed by earthquake-induced landslides during earthquake emergency rescues. *Earthquake Research in China, 25*(3), 333–342.
Wasowski, J., Pierri, V., Pierri, P., & Capolongo, D. (2002). Factors controlling seismic susceptibility of the Sele Valley slopes: The case of the 1980 Irpinia earthquake re-examined. *Surveys in Geophysics, 23*(6), 563–593.
Wilkinson, S., Grant, D., Williams, E., Paganoni, S., Fraser, S., Boon, D., et al. (2013). Observations and implications of damage from the magnitude Mw 6.3 Christchurch, New Zealand earthquake of 22 February 2011. *Bulletin of Earthquake Engineering, 11*(1), 107–140.
Wilson, R. C., & Keefer, D. K. (1983). Dynamic analysis of a slope failure from the 6 August 1979 Coyote lake, California, earthquake. *Bulletin of the Seismological Society of America, 73* (3), 863–877.
Yagi, H., Yamasaki, T., & Atsumi, M. (2007). GIS analysis on geomorphological features and soil mechanical implication of landslides caused by 2004 Niigata Chuetsu earthquake. *Journal of the Japan Landslide Society, 43,* 294–360.
Yagi, H., Sato, G., Higaki, D., Yamamoto, M., & Yamasaki, T. (2009). Distribution and characteristics of landslides induced by the Iwate-Miyagi Nairiku earthquake in 2008 in Tohoku district. *Northeast Japan. Landslides, 6*(4), 335–344.
Yamada, M., Wang, G., & Mukai, K. (2013). The classification and features of earthquake-induced landslides in the world. In *Earthquake-Induced Landslides* (pp. 117–124). Springer.

MIX
Papier aus verantwortungsvollen Quellen
Paper from responsible sources
FSC® C105338

If you have any concerns about our products,
you can contact us on
ProductSafety@springernature.com

In case Publisher is established outside the EU,
the EU authorized representative is:
Springer Nature Customer Service Center GmbH
Europaplatz 3, 69115 Heidelberg, Germany

Printed by Libri Plureos GmbH
in Hamburg, Germany